BACKROADS & BYWAYS OF

UTAH

BACKROADS & BYWAYS OF
UTAH

Drives, Day Trips
& Weekend Excursions

Christine Balaz

The Countryman Press
Woodstock, Vermont

We welcome your comments and suggestions.

Please contact

Editor
The Countryman Press
P.O. Box 748
Woodstock, VT 05091

or e-mail countrymanpress@wwnorton.com.

Backroads & Byways of Utah
ISBN 978-0-88150-906-9

Book design by Hespenheide Design
Map by Erin Greb Cartography, © The Countryman Press
Interior photos by the author unless otherwise specified
Composition by Chelsea Cloeter

Published by The Countryman Press, P.O. Box 748, Woodstock, VT 05091

Distributed by W. W. Norton & Company, Inc., 500 Fifth Avenue, New York, NY 10110

Printed in the United States of America

10 9 8 7 6 5 4 3 2 1

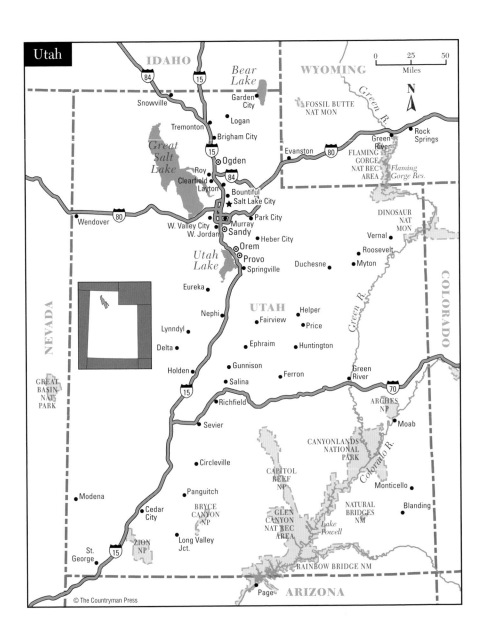

Utah

IDAHO

Bear Lake

WYOMING

0   25   50
Miles

N

84
15

Snowville

Garden City

Logan

FOSSIL BUTTE NAT MON

Green R.

Tremonton

Brigham City

Green River

Rock Springs

Great Salt Lake

15

Ogden

Evanston

80

Roy

84

Clearfield

Layton

Bountiful

Salt Lake City

Wendover

80

W. Valley City

W. Jordan

Murray

Sandy

Park City

Heber City

FLAMING GORGE NAT REC AREA

Flaming Gorge Res.

DINOSAUR NAT MON

Vernal

Roosevelt

Orem

Provo

Springville

Utah Lake

Duchesne

Myton

Eureka

Nephi

Fairview

UTAH

Helper

Price

Lynndyl

Ephraim

Huntington

Delta

Holden

Gunnison

Ferron

Green River

Salina

70

Richfield

ARCHES NP

Green R.

Colorado R.

NEVADA

15

Sevier

Moab

CANYONLANDS NATIONAL PARK

GREAT BASIN NAT PARK

Circleville

CAPITOL REEF NP

Modena

Panguitch

BRYCE CANYON NP

Cedar City

Long Valley Jct.

ZION NP

15

St. George

GLEN CANYON NAT REC AREA

Lake Powell

NATURAL BRIDGES NM

Monticello

Blanding

Colorado R.

RAINBOW BRIDGE NM

Page

ARIZONA

COLORADO

© The Countryman Press

Contents

Introduction

Utah boasts vast amounts of open land and extremely varied terrain. It has mountain peaks that stand more than 13,700 feet above sea level, broad valleys, high planes, and many varieties of desert: red rocks, vermillion cliffs, buffed sandstone canyons, rounded arches, lunar landscapes, dramatic gorges, and expansive salt flats. Within a few hours of each other are high, fortified, alpine highways and wild dirt roads stretching across broad, low-lying desert basins. To drive for an afternoon in this state is to visit more landscapes than you might see driving for days in other parts of the country. To experience all of them is to travel to nearly every continent in the world—except, on most days, Antarctica—flora-, fauna-, and geology-wise. And Utah has six national parks and numerous national monuments and state parks to celebrate and preserve these vistas.

Because roughly 75 percent of Utah's population lives along the **Wasatch Range,** the remainder of the state's lands is quite open and untainted by the trappings of civilization. The general openness of the landscape and straightness of the roads enable relatively speedy long-distance travel. Thus, many of the drives in this book will be in excess of 100 miles. The list of suggested routes will begin with a tour of the Wasatch canyons near Salt Lake City. The rest is then listed in clockwise order, beginning in the northeastern corner of the state and working around to many remote and gorgeous sections of road. Each route has been chosen to represent the most interesting, beautiful, visually diverse, and historically

Aspens High on Mt. Timpanogos' Shoulder

and culturally relevant sections of road and points of interest flanking these.

Each drive listed in this book has been embellished with the most interesting historical, geological, scenic, and factual information possible. Keep your eyes open for advisories that warn about climate-related hazards such as potential for snow closures or excess heat, as well as other possible dangers such as long stretches of road without services.

Utah is a large state—the 13th largest in the union—with only one major urban population, which runs the western Wasatch Front. Nearly everyone traveling to Utah by plane will almost assuredly fly into **Salt Lake City** via the **Salt Lake International Airport,** a major hub just five min-

UTAH HISTORY

The last major ice age, the Wisconsin Glaciation, lasted from about 70,000 B.C. to 10,000 B.C. During that time, the world was a much cooler, wetter place. The Great Basin, a massive bowl centered around, but bigger than, Nevada has no outlet to the ocean; during this wet time, it became a huge holding area for water. Several times a massive lake in northern Utah (and more than a quarter the size of the state) filled, drained, and refilled. Its last major stint ended about 12,000 years ago when a weakness along its shores gave way, releasing **Lake Bonneville** in a catastrophic flood across southern Idaho. Prior to the dam breaking, the lake's waters would have been up to 800 feet over present-day Salt Lake City.

It wasn't until after the ice ages started to recede that Utah became inhabited by humans. In fact, right around 11,000 years ago, the first groups of people started leaving evidence of their existence. Today this evidence has been found across the state, with the oldest being in **Danger Cave** (see chapter 15), in the **Bonneville Desert.** These most ancient residents lived a transient lifestyle, by appearances spending all waking hours just trying to stay alive. The first permanent peoples in Utah to live in larger groups and leave major traces were the Anasazi and Fremont. Partially nomadic, these people also planted basic crops and even raised animals for consumption. This relatively stationary lifestyle allowed for the manifestation of more advanced culture, such as tool and basket making. These were the ancestors of, and would later evolve into, the **Ute, Paiute, Shoshone, Bannock,** and **Gosiute.**

Far from any coasts, Utah was quite late being inhabited by Europeans. Some of the first Euros to see the region were the Spanish explorer-priests, **Dominguez** and **Escalante** (after whom many aspects of Utah's southern region are named). These two, stationed in Sante Fe, came into the area on an exploratory mission to open a direct passageway from south to north in 1776. A particularly cruel winter turned these men around, but not before they had come well into northern Utah—near **Dinosaur National Monument** and the **Uinta Basin.** An early winter swept in and ushered the party hastily back south. After this expedition came the 18th-century Spanish miners, whose presence in the **Uinta Mountains** is evidenced by remaining infrastructure, tools, and the altered earth they left behind.

The next known Europeans to enter the area were all itinerant in nature: trappers, traders, and other passers-through. Among the roster were **Kit Carson** and **Antoine Robidoux,** who established self-named trading posts in the 1830s and

'40s, **William Ashley,** namesake of the Ashley National Forest, and **Etienne Provost,** whose name appears in numerous places across the state. Settlement beyond these trading posts was nonexistent.

One of the earliest and most influential explorers of the Intermountain West, **John Wesley Powell,** ventured into the region via the **Green** and **Colorado Rivers** on two separate occasions. At the time of his voyages (1869 and 1871), white-water boating wasn't even a concept—and these rivers served up some very serious rapids. On his trips, Powell scrupulously observed and recorded the West's enormous and otherwise uncharted landscape. In addition to the geology and ecology of the area, Powell actually detailed this new white-water phenomenon in his immaculate journals. Not without accidents and losses, his crews were nevertheless mainly successful—led by their one-armed captain.

Finally in 1848, Utah saw its first large, permanent settlements—facilitated by the arrival of Latter-day Saints in Salt Lake Valley in July of that year. Despite demeaning rumors accusing the Mormons of mistaking the Great Salt Lake for the Pacific Ocean, they actually had specifically planned and prepared to settle in the Great Basin. Suffering persecution wherever they lived, the Mormons desired a home completely isolated from society. So with this specific wish, they set about establishing an empire in Utah, hoping to even create a sovereign nation of their own. Though they did a remarkable job of embedding a network into the state and beyond, they were never able to create their own country.

---

utes west of downtown. Be prepared: In the West, people are accustomed to driving long distances. And since Utah is crisscrossed by several major interstates, and speed limits are relatively high, most people consider a 200 or 300 mile drive totally reasonable for a weekend trip.

Though Salt Lake City is certainly worth visiting, it will not be included in this book. (If you are interested in reading more about this city—or other cities in Utah—on your journey, you may want to check out Countryman's *Great Destinations: Salt Lake City, Park City, Provo & Utah's High Country Resorts* or *Utah: An Explorer's Guide.*) With regards to rural exploration across the state, read the tips and alerts below:

National Parks Pass
Utah has a bounty of National Parks and Monuments. In fact, it has 13 in all—or 17, if you count the four national trails—with yet more abutting its borders. For those planning to spend any serious amount of time in the

Mouth of Kolob Canyon

state, it is almost assuredly worth the money to purchase an annual pass. Formerly known as the Golden Eagle Pass—now the **America the Beautiful National Parks and Federal Recreational Lands Pass**—it grants unlimited access to all national parks and monuments (as well as some other government-run parks) for a year from date of purchase for $80. Compared with the cost of one-week passes to individual parks (usually around $20 a piece), this could very well be worth the money—especially for those planning to visit other under-the-radar types of federally operated destinations such as **American Fork Canyon** near Lehi (chapter 3 in this book) or the **Mirror Lake Highway** in the **Uinta Mountains** (chapter 5).

(Lack of) Cellular Phone Service

Be aware that in Utah cellular phones are often completely without service. Long distances between townships and mountainous topography, combined with sparse rural population (and therefore low demand for cellular technology), renders predictably bad-to-nonexistent backroads cell service. If heading well off the beaten path, it doesn't hurt to occasionally check the phone for service, making a mental note of the last point on the approach where the phone picked up solid reception. For people accustomed to readily available phone use, Utah's lack of cell service can be a real shock. Keep in mind that it can potentially amplify the isolation should something go wrong in rugged and remote situations.

Natural Hazards

Where there's natural beauty, there's potential for danger. Utah's grandeur is typified by trademark ruggedness, barrenness, isolation, and other flavors of inhospitableness. While these create beautiful vistas, these features and landscapes are not the type to foster life—particularly untrained human life perhaps out of touch with man's vulnerability in nature. All of these potential hazards, of course, are amplified exponentially by frequent lack of human company, cell phone service, and rescue resources.

Water: Too Little

Ranking high in Utah's biggest threats is its polarized and fickle climate: extreme temperatures and the lack (or sudden surplus) of water. Lack of water is, by far, the region's most common threat to the unprepared—particularly in southern Utah. To prevent dehydration, always pack extra water in the car (or in the backpack, if hiking). Depending on the season and strain of the activity, it is not excessive to bring *at least* a gallon of water per person, per day.

Water: Too Much

Considering the above advisory, it's strange to find that *too much water* can be a threat in Utah as well—yet each year people fall victim to flash flooding in this state. These floods generally occur in narrow canyons (slot canyons and other narrow river canyons) and are especially hazardous, as they often take their victims by complete surprise. Most accidents occur when people are recreating off the grid. Though it may be sunny and cloudless in their immediate vicinity, floods can form as a result of distant storms. These can easily inundate a drainage system and send a wall of rushing water downstream with no warning, packed with debris that will kill. Your best precaution is to meticulously check the weather before spending any time in a canyon system and to stay far clear of any slot canyons if there is rain forecasted anywhere in the region.

The other surplus-water-related threat, though much less violent, is mud. Some of Utah's soil, when mixed with rain, can create roads so slick or sticky that they become impassable for days at a time, stranding those on the "other" side from civilization. Especially if exploring Utah's old lake beds and salt flats, be sure to monitor the weather. Though a road may appear dry, more than a foot of car-sticking mud could be lurking just beneath the surface, ready to eat a passing car. Extraction from such a mess requires the payment of a hefty fee to a fancy, military-grade tow truck operator, as most typical wreckers cannot do this type of job!

Sun

Sunblock. Sunscreen. Hat. Protective clothing. All things Mother suggested will be of indescribable use here! Utah, regardless of season, can burn the skin senseless. Summer heat means bodies are scantily clothed and exposed. Winter snow forms a blanket of white on the ground that acts as a mirror and absolutely blasts skin with ultra violet rays. Add to that the state's relatively high elevation, roughly 300 days of sun a year, and we've got premature aging. Be prepared!

Mountainous Terrain

Dramatic topography can be in and of itself a threat to the unwise. The fact is that gravity exists 24 hours a day, seven days a week—and not all ground is stable. Be cautious near long drop-offs, even if when walking on a trail. And be mindful when walking off-trail. Much of Utah's wilderness consists of fragile desert or high-alpine terrain easily damaged by foot traffic. Additionally, complicated mountain topography can easily render even experienced hikers lost. Be careful to keep your bearings and be diligent about returning exactly along the approach path as you came.

Lions and Tigers and Bears: Uh-Oh

Animals, too, can present a lethal danger. It's obvious that mountain lions and bears are a threat. (But don't worry; tigers only exist in zoos in Utah.) However, don't forget the smaller (snakes and spiders) and the less-obvious threats (moose and mice). Some underestimate the potential danger of black bears, assuming they're less dangerous than grizzly bears, but American Fork Canyon had its first lethal encounter in 2006. In that particular case, food was left very near the sleeping victim and not properly tied off far from the campsite.

Moose, too, are highly dangerous, testy, and unpredictable. They appear dumb, lanky, and homely—but they have a mean temper, particularly around their offspring, and can easily trample a person. Snakes exist in plentiful amounts in Utah; it's got the perfect climate. Be particularly on the lookout for rattlesnakes. They are the most lethal of the state's serpents. But, fortunately, the rattle affixed to this snake usually serves as fair and generally sufficient warning to back away. Upon hearing this, simply pause,

Rattlesnake in the West Desert's House Range

locate the sounding snake, and calmly (or with as much calm as can be mustered) leave its vicinity.

Finally, mice carry hantavirus. It is excreted in their waste and saliva, and can easily become airborne if their droppings are disturbed. If particles of infected waste are inhaled, the disease can be caught by unlucky humans. Animal danger should not keep people out of the wilderness, but wildlife should be treated with respect (whether or not they pose a human threat). Give them space to operate without harassment, and don't tempt them into destructive habits.

Vehicle Awareness

As almost all of these drives leave cell service at some point (and some for the majority of the route), drivers are responsible to make sure their vehicle will last the drive. Depleting the gas tank, flatting a tire, and running too low on oil are easy problems to prevent, but difficult to fix when in the backcountry. In the event something does go awry, it's best to have extra clothing, food, and clothes in the car.

All of that said, Utah is one of the nation's most beautiful and rewarding states to visit. Don't shy away from its adventures. Simply come prepared to be responsible for your own wellbeing.

Backcountry Skier in Little Cottonwood Canyon's White Pine Basin

CHAPTER

1

Salt Lake City into Little Cottonwood Canyon

Estimated length: About 25 miles one-way from downtown Salt Lake City to the top of Little Cottonwood Canyon

Estimated time: A bit more than 30 minutes' pure driving to Alta; a short half day for a round-trip journey with a few stops

Getting There: Thanks to the I-215 beltway encircling Salt Lake City, the Cottonwood Canyons couldn't be easier or quicker to reach. From the southeastern corner of this square-shaped interstate loop, take exit 6, and head east on UT 190. This highway climbs eastward and uphill, turning naturally into South Wasatch Boulevard as it dips to the south. Stay on this road as it heads south along the base of the Wasatch for just more than 2 miles. Turn left, following signs, to head uphill on UT 210 and into the very obvious, ascending canyon. (Starting near the appropriate I-215 exit, the route to the canyons is signed the entire way; Big Cottonwood is the home of Brighton and Solitude; to get to Little Cottonwood, follow signs to Alta and Snowbird.)

Highlights: This trip is a short and sweet jaunt up Little Cottonwood Canyon. But being so near to Salt Lake City, quite beautiful, and endowed with a bit of interesting history—as well as phenomenal modern recreation and dining—it's worth the excursion. Along the way is the mysterious-seeming subterranean **Mormon vault, Snowbird Resort,** the "Hellgate," and **Alta Lifts.** Spring through fall, this canyon opens itself up to rock climbers and boulderers, road bikers, and hiker. In winter, this canyon fills with skiers and snowboarders (skiers only at Alta!). The resorts, also open in summer, offer hiking trails, summer lift and tram rides, as well as restaurants and pubs. Occasional live music and other festivals take place during the peak of summer.

Done in conjunction with the **Big Cottonwood Canyon** route (chapter 2 of this book), this route will amaze the viewer with geological contrast. Pristinely U-shaped, and very smooth, it's not difficult to imagine glaciers polishing out the white granite-like rock of Little; angular and blocky, it would appear that Big Cottonwood belongs in another mountain range entirely.

Before even entering Little Cottonwood Canyon, the driver must first ascend to its base! The **Salt Lake Valley,** at around 4,300 feet elevation, sits the better part of a vertical mile beneath the *base area* of Alta (8,950 feet) at the end of the road. Not to mention that the tip-top of Snowbird, **Hidden Peak,** has an elevation of 11,000 feet. Clearly glacially sculpted, the white walls of this canyon are actually not granite. Commonly mistaken, they actually are composed of quart monzonite—also an igneous, intrusive rock, its quartz composition is roughly 5 to less than 20 percent by mass (whereas granite contains 20 percent). Remember that, and you'll know more than most people in Salt Lake City about the canyon's geology. In

---

NO DOGS ALLOWED!

Little Cottonwood Canyon and Big Cottonwood Canyon are both part of Salt Lake City's municipal watersheds. Because of this, and the already high use of these canyons, the Salt Lake City government strictly forbids the presence of dogs in these canyons, regardless of leash use. There are very few exceptions to the rule, and the penalty for disobeying can be a fine of several hundred dollars and, in extreme cases, can include jail time. Those with pets should leave them at home; even dogs left in parked cars are considered illegal. Alternative, dog-friendly canyons are **Mill Creek Canyon,** east of Salt Lake City, at the end of 3800 South Street, and **City Creek Canyon,** north of the city's **Capitol Hill.** (In City Creek, dogs are only permitted downstream of the obvious water treatment center a few miles up the canyon).

---

winter, be sure to call the road report (801-536-5778), especially after a big snowstorm, to make sure the highway up canyon is in fact open; with such steep sides, and such a narrow bottom, the canyon must be closed often for preventative blasting and clearing.

This rock was quarried by the Latter-day Saints (LDS) Church for use in the **Salt Lake Temple, Utah Capitol Building,** and **LDS Office Building.** Not actually mining from the within the ground, they simply blasted and (cleanly) broke up already-fallen boulders. A great dismay to rock climbers, this has destroyed a lot of the classic boulder problems in the canyon. This LDS quarrying largely took place near the mouth of the canyon, but farther up as well. Parking at any of the numerous shoulder pullouts (or the park-and-ride at the mouth of the canyon, used heavily in winter by **Utah Transit Authority** ski shuttle buses), and walking even just slightly into the scrub oak forests will reveal many boulders with obvious blasting marks.

Even the Mormons have mistaken Little's rock type in the name of their **Granite Mountain Records Vault.** This obvious, large hole in the rock about a mile into the canyon on its northern side is a deep, underground storage facility owned by the LDS Church and is heavily guarded. Created for the safekeeping of church documents in the event of an apocalyptic event, the vault also contains more than 35 billion genealogical files mostly on microfiche—a collection that was started back in the 1930s.

Needing a secure place to store these temperature-sensitive records, the church discussed many possible locations, including downtown buildings and other canyons. But the solid quartz monzonite walls of Little seemed a perfect and safe place. After initial testing, excavating began in 1960 using diesel fuel and ammonium sulfate. Ten-foot sections were blasted at a time, with alternate days being used to remove the fallen stone. The tunnels were completed enough by 1963 to begin storing items. The structure consists of three, nearly 700-foot-long tunnels with a few cross-tunnels. The doors to the vaults are allegedly capable of sustaining a nuclear blast. Not too unlike something from the Twilight Zone! Today private persons can pay a pretty penny to have their own goods stored in the vault as well.

On the way up the canyon, several hiking trails depart from obvious pullouts on either side of the road. Most of these head up the southern side of the canyon, starting several miles up in the canyon where walls become less steep and rocky; if they started closer to the mouth of the canyon, the hikes would dead-end in cliffs and steep talus piles. A trail departing to **Red Pine Lake** and **White Pine Lake** beings at a rather large parking area (on the south side of the road) 5.5 miles up canyon. The hike to Red Pine Lake from here is about 3 steep miles, gaining almost 2,000 feet of elevation in that short distance. The White Pine Trail is just an extended version of this, lasting 4.5 miles each way, and gaining 2,540 feet in elevation. At the parking lot, all of this information is clearly spelled out on a kiosk map. Hiking into either lake makes it obvious that the Wasatch Range is a huge area of wildlands. These trails are used by rock climbers in summer to access yet higher alpine climbs; in winter, skiers and snowboarders ride down them, particularly White Pine, which can be accessed from Snowbird with just a bit of hiking.

Snowbird is about 8 miles up Little Cottonwood Canyon and is the first resort encountered on the drive up. Most famous as a ski and snowboard resort, Snowbird gets more than 500 inches of snow each winter. One of Utah's biggest, and certainly one of its most "hardcore" areas, it has 2,500 acres, 3,240 feet of vertical drop, and a base-to-summit tram that summits

---

UTAH'S OWN ALPE D'HUEZ

Anyone who's watched **Le Tour de France** knows about the famous mountain stage called **l'Alpe d'Huez**. Perhaps the most legendary and classic mountain stage on the world's most prominent bike race, the mountain has a nearly identical profile (in terms of distance and grade) to that of Utah's own Little Cottonwood Canyon. Both are just about 8.5 miles in length, climbing about 3,500 feet, with an average pitch of around 9.2 percent. The only difference is that the French version has switchbacks, and the Little Cottonwood version is more or less a straight shot, with some curves being the only thing that hides the ride and keeps the rider from giving up. And they're an ocean apart.

A tip for road bikers: this canyon is one of the Wasatch's more beautiful, challenging canyon rides. Yet, with a fair amount of traffic on a pretty narrow roadway, this can be quite unpleasant during the day. Also, such a steep descent causes many drivers to burn their brakes. Not as terrible as another driver, the pungent smell of burning brakes can make sucking wind on a hill climb quite undesirable. The perfect solution to this problem is to wake up extra early and start riding by 5:00 AM. Riding at this time, in the middle of summer, means avoiding almost all of the vehicular traffic. It also provides ideal temperatures. Mid-day summer riding in this canyon can be pretty sadistic; its bowl-shaped bottom bakes in the sun and can be a cooker.

Ridgeline between Red Pine and White Pine Lakes in Little Cottonwood

in just 10 quick-seeming minutes via a 1.6-mile cable. This tram also operates during the summer, and for a few dollars brings those unable to hike (or unwilling) up to an incredible view. Those who do hike to the top can ride the tram down for free, saving their knees for another hike! During the winter, the resort is absolutely hopping, but in summer its restaurants remain open—from fine, American and Continental dining at the **Aerie** (in the **Cliff Lodge**), to basement and patio bars serving beers and burgers. Many, many restaurants and hotels operate at Snowbird; those curious should investigate further by calling the resort or checking online.

The **Cliff Lodge** is Snowbird's most obvious and large hotel, with 11 stories (and a glass wall entirely exposed in its floor-to-ceiling atrium), three restaurants, multiple pools, and an enormous oriental rug collection. Evidence of the 1989 World Cup climbing competition is still visible on its external walls, in the form of still-attatched holds. On its 10th floor is the **Aerie,** Snowbird's finest dining. With an incredible wine list (more than 850 selections), almost floor-to-ceiling windows, incredible down-canyon and mountainside views, and a high-caliber, American and Continental menu, this is among the finer restaurants in the Salt Lake City area. And for those who don't know, this town is actually very well endowed with culinary arts from around the globe. A sushi menu is available here as well. Those without reservations or who are looking for a quicker meal can sit at the sushi bar, just outside of the main restaurant.

Rock Climber on the Northern Side of Little Cottonwood

Moving uphill just slightly is **Alta Lifts.** Deliberately avoiding the "resort" moniker, Alta prefers to reserve itself for those who appreciate the less fast-paced aspects of skiing. Today (and likely forever), snowboarders are not allowed at Alta—unless they are skiing. Though not as big or as flashy as its neighbor, Snowbird, Alta is more beloved by many. It is a mountain characterized by semi-secret stashes that must be hiked to, chutes and bowls, as well as tree runs that are all divided up from each other.

In the summer, this sloop-and-bowl mountain terrain holds a collection of small, alpine lakes. Intermixed forest and fields sit in Albion Basin beneath Devil's Castle, a massive limestone cliff that actually has multiple rock climbs reach all the way to its peak (as well as a non-technical hiking trail that reaches its peak via the shoulder). Several trails crisscross the area, making almost all of it easily accessible on foot and by mountain bike.

Wildflowers bloom intensely here in the summer, giving cause to the Alta installment of the Wasatch **Wildflower Festival.** This three-day weekend affair brings many people to walk these slopes when they are absolutely painted with vibrant flowers of all colors. (Just one of the days takes place at Alta; the others occur in Big Cottonwood Canyon, at Brighton and Solitude.) Albion Basin has more than 300 species of wildflowers, and because of its high altitude, they don't come into full bloom until right in the middle of summer. The festival features guided walks, food, and live music. Hikes of varying difficulties are offered, so not everyone must be of a high fitness level.

Alta, as with many, many ski resorts across the western United States, was originally a mining town. In its heyday during the 1860s and '70s, it had flourishing mining operations, supported by a community with 26 bars, 6 breweries, and uncounted other non-pious goings on. Notice the name "Hellgate" around Alta: On the way up the canyon, immediately before Alta, two giant limestone fins come down and almost meet, nearly pinching off the canyon at that point. The Mormons named this section of Little Cottonwood, "Hellgate," as if it were literally the gates to Alta's "satanic," beer-drinking, mining lifestyle. Like most mining towns, Alta died quickly when its limited resources ran out and left its residents stranded high above Salt Lake City with no viable economy. Also, as with many such ghost towns, Alta saw a revival as a ski resort. Alta actually became one of Utah's first ski areas when it opened in 1938 (just behind Brighton, which opened in 1936). Since then, the facility has obviously grown, but it has maintained a stiff arm's length from large-scale resort business practices.

Those interested in the mining history of the canyon can check out **Grizzly Gulch** via a hike. Along the way, this crosses many tailings piles and random mining equipment. A 2.2-mile walk reaches the **Prince of Wales Mine Shaft.** In operation from 1872 through the Great Depression, this mine shaft, was roughly 900 feet deep, leading down to a maze of tunnels. Today a metal grate sits over the shaft, lest unsuspecting backcountry skiers plunge into it. The original hoisting machine still sits there today. To get there, park 8.25 miles up the canyon, at the **Old Forest Service Garage.** A road leading up behind the Alta City Building turns into a jeep road, and eventually a trail that steadily climbs into the gulch.

Alta marks the end of the route. To return, simply downshift (to save the brakes) and head back toward Salt Lake City. Again—a road more than 8 miles long with an average grade of 9.2 percent should be taken slowly. Remember the burnt brake smell on the way up? It is very possible to link this drive with **Big Cottonwood Canyon** (chapter 3 of this book), which runs parallel to, and just north of, Little on UT 190. Right next to Little, but a much different canyon geologically, Big is the home to **Brighton** and **Solitude** ski resorts.

IN THE AREA

Accommodations

Cliff Lodge, Snowbird 84092. Moderately priced. Snowbird's largest hotel with guest rooms, spa, pools, and three restaurants. Call 1-800-232-9542. Web site: www.snowbird.com.

Attractions and Recreation

Alta Ski Lifts Company, UT 210, Little Cottonwoods Canyon, Alta 84092. Ski area with summer activities and hiking. Call 801-359-1078; canyons road report: 801-536-5778. Web site: www.alta.com.

Snowbird, UT 210, Little Cottonwood Canyon, Snowbird 84092. Ski and snowbird resort in winter, summer resort in summer (with tram rides to resort top). Many restaurants and pubs. Call 801-933-2222; canyons road report: 801-536-5778. Web site: www.snowbird.com.

Dining

Aerie Restaurant, Cliff Lodge, Snowbird 84092. Expensive. Open Tuesday through Saturday for dinner only; reservations strongly recommended. American and Continental cuisine, with a sushi bar. Call 801-933-2181. Web site: www.snowbird.com.

Other Contacts

Utah Transit Authority. Operating winter ski shuttle buses running from the parking lot at the mouth of Little Cottonwood Canyon (in addi-

Autumn Hits Little Cottonwood

tion to standard metro bus lines). Call 801-287-2667. Web site: www.uta
.com.

Wasatch Wildflower Festival. Taking place for three days (only one of
which is held at Alta; the others feature the flowers of Brighton and Solitude resorts in Big Cottonwood Canyon) in the middle of July. Call Alta
Visitors Bureau at 1-888-258-2840. Web sites: www.discoveralta.com and
www.wasatchwildflowerfestival.org.

Ski Touring atop Reynold's Pass, Big Cottonwood Canyon

2

Salt Lake City into Big Cottonwood Canyon

Estimated length: About 30 miles one way from downtown Salt Lake City; add about 11 miles to descend into Park City over Guardsman Pass (open in summer only)

Estimated time: Just more than half an hour as a straight shot to the top of Big Cottonwood Canyon (or an hour to reach Park City); about a half day with a few stops and a quick hike

Getting There: The I-215 beltway that encircles the expansive Salt Lake metropolitan area makes it possible that almost every place in Salt Lake City is within 20 minutes of the mouth of Big Cottonwood Canyon, barring major traffic. to Big Cottonwood by departing from the far southeastern corner of this beltway, taking exit 6, and heading east and uphill on Fort Union Boulevard/UT 190. This climbs eastward and uphill, turning naturally into Wasatch Boulevard as it heads south. At a very obvious junction, signs will point eastward into Big Cottonwood Canyon (UT 190) and toward Brighton and Solitude, prompting a left-hand turn. (Do not stay straight here or follow signs to Alta and Snowbird; those are in Little Cottonwood Canyon, which runs parallel to Big, a few miles farther south.)

In winter, be sure to call ahead of time for a road report (801-536-5778). Especially after big snowstorms, this highway is frequently closed for up to several hours; with such steep sides, and such a narrow bottom, the authorities must perform preventative blasting and clearing.

Highlights: Much like its sister, Little Cottonwood Canyon, Big is the home of two ski areas, a lot of hiking, and a very steep road prefect for biking. Also like Little, Big Cottonwood is part of the city's watershed and is a no-pets-allowed area. However the road to the top of Big is quite a lot longer than that in Little; it stretches for almost 15 miles (instead of Little's 8.3). Furthermore, at the top of Big, the road does not end (unlike in Little); it actually continues up and over the ridge line at **Guardsmans Pass,** descending into **Park City,** arriving in **Deer Valley**—but don't expect to take this road in winter. A dirt road at the top, it is closed during the long, snowy season. Along the way, the route is flanked with the **Mount Olympus and Twin Peaks Wilderness** and the numerous foot trails that access them. By definition, these are not accessible by vehicle, but they provide pristine environments in which to ski tour and hike.

In summer, **Big Cottonwood Canyon** is one of the most popular hiking destinations for Salt Lake City-ites. In winter, its two ski resorts, **Brighton** and **Solitude,** massive backcountry, and annual 500 inches of snowfall make it a world-class ski playground. Though it would seem ages away, Park City and its ski hills are actually just over the hill from Big Cotton-wood as the crow flies. Many backcountry skiers take advantage of this fact, often riding the Canyons Resort's lifts and then embarking on a major ski tour into and through Big Cottonwood Canyon.

The actual history of Big Cottonwood dates back roughly 900 million years—to the time when its quartzite and slate rock formations formed. Both metamorphic rock types, its rust-colored quartzite originated as sandstone, while the black slate came from shale. Compressed under high pressure, both turned (or "metamorphosed") into these more compact, but visually similar, versions of their originals. Especially low in the canyon, this orangish-red and very blocky, striated quartzite is the obvious and domi-nate rock type.

Just as in Park City and Little Cottonwood Canyon—and all across Idaho, Wyoming, and Colorado—Big Cottonwood's ski resorts are a replacement industry for earlier mining towns. In 1863 the first mine in the canyon, Evening Star, was opened. In the 1870s and '80s, more than 2,400 miners would come to live up near Brighton and Solitude, working at lit-

Spring snow high in Big Cottonwood

erally hundreds of assorted mines, living geographically and culturally seg-
regated from the young, very Mormon Salt Lake City below. These towns
flourished for just a few decades, then died out as quickly as they appeared.
Same as in other mining towns, the resources here were finite, and this
isolated settlement had no recourse once the mines became obsolete. Massive timber harvesting took place during the intervening years, but the
canyon's local economy would not properly rekindle until Brighton opened
in 1936, followed by Solitude in 1957.

Since human settlement in Salt Lake Valley, Big Cottonwood has provided a beautiful retreat from city life—from picnic grounds, to long,
mountainous hikes, these canyon's resources are spread out, yet quite near
the city. Today, this is the most popular sport-climbing area immediately
near Salt Lake City, and it is one of its best backcountry skiing zones. When
getting in the car, remember to leave the dog at home. Even if a dog remains
in the vehicle the entire time, its presence is still highly illegal in Big Cot-

The Quartzite Climbing Crags above Big Cottonwood Canyon's S-Curves
Ron Winsett

tonwood. Law enforcement is serious about the preservation of the city's watersheds and will happily award a several hundred dollar fine (and possibly even jail stint) to anyone disobeying this simple law.

Porcupine Pub has the best location for people traveling to Little or Big Cottonwood canyons. A large, nice-but-casual, restaurant and pub, this popular joint serves draft beer, massive (and delicious, heaping) nachos, and a full menu that is perfectly befitted to skiing, hiking, and mountain biking appetites. Burgers, steaks, salads, and other entrees top the menu, with great daily specials offered. Located right at the southwestern corner Fort Union Boulevard and Big Cottonwood Canyon Road junction, this is one of the last stops to be had before entering the canyon. Those who miss it on the way up should hit it on the way down.

Storm Mountain Picnic Site is the first major recreation area, located just 2.75 miles up canyon. R. D. Maxfield Jr. decided to make his home here in the early 20th century—and this has been a leisure site for many decades, with its stage being built by the Civilian Conservation Corps in the 1930s. Today, this popular day-use spot has plenty of open, streamside space, a small outdoor stage, and a good bit of rock climbing, too. To enter, a day fee must be paid. (Note: no drinking water is available here.)

Four and a half miles into the canyon is the extremely obvious "S Curve" of the road. Tucked in the folds of this curve (as well as laying to the outside of the curve) are a few dozen cars' worth of parking spots. This is a heavy recreation-use spot, again with a generous amount of climbing (located above the upper fold of the curve), and a few hiking trailheads. On the lower bend of the S-Curve is one of Big Cottonwood Canyon's most well known hiking trailheads, called **Mill B.**

A 3-mile trail to **Lake Blanche** departs from here. This stunning alpine lake is encircled by rock spires skirted by talus and smattered thinly with evergreen trees. **Sundial Peak,** elevation 10,320 feet, sits immediately near the lake, with other summits surrounding it in the distance—**Dromedary Peak** (11,107 feet), **Sunrise Peak** (11,275 feet), and **Monte Cristo** (11,132 feet). Very dissimilar in appearance from the canyon bottom, the gray stone of this basin has been polished smooth by ancient glaciers in many places. To get there, head east (up canyon) out of the lower parking lot in the S-Curve (Mill B). Go about 0.25 miles on the asphalt trail to a bridge that crosses a creek. The trail to Lake Blanche (now dirt) is signed, forks up from here, and is very easy to follow. The remainder of the hike is just under 3 miles. Before reaching Lake Blanch (located at the terminus of the trail), other side trails will head in quick jaunts to each Lake Lillian and Lake Florence—both fine-looking themselves.

For an almost zero-walking-time alternative, leave the Mill B parking lot and carefully cross the S-folds of the road to the north, toward a small alcove and stream. Hardly uphill from the road at all, the obvious stream forms a waterfall. Hiking along the trail (that heads uphill and to the right), leads to a few tiers of cliffs; these are some of the most popular rock climbing walls in the canyon. Beware, though—the approach to these is on steep, loose talus, and can be dangerous to walk for those not accustomed to such scrambling.

Continue uphill, undistracted by infrastructure of any kind, and simply enjoy the scenery. Consider taking a hike along the way, as nearly each bend in the road reveals another pullout with another trailhead. Given enough time, these can be explored either with or without a map. During the winter, many of these serve as excellent backcountry skiing access points.

Almost at the top of the canyon are Brighton and Solitude resorts, Solitude being the lower of the two and having an apt name. Though "just" 1,200 acres, it has a nearly 2,050-foot drop, and hardly any crowding whatsoever. Perhaps because of its "old-fashioned" lifts, or the fact that its surrounding resorts are more famous, it enjoys seemingly less congestion than

---

RIDE BIG COTTONWOOD

Big Cottonwood Canyon is just another of the Salt Lake area's great canyon road rides. This is its most serious single hill climb, too, with an in-canyon distance of 14.7 miles to Solitude and an elevation gain of 4,584 feet. And that's not counting the initial climb from the valley to the canyon's mouth, which adds another 600 or 700 vertical feet, depending on the point of origin.

Those wanting to take it even further can top out at **Guardsman Pass** (the turnoff to which is just slightly before Solitude Resort), by continuing on the switchbacks, to an elevation of 9,700 feet (and a total distance of 16.9 miles), with final grades as steep at 15 percent. The reason for this not typically being part of the ride profile, is that this topmost section of road eventually turns to dirt, and many don't want to ride their multi-thousand-dollar bikes on dirt. For more information on this canyon, or other rides in the area, check out www.saltlakecycling.com. This awesome Web site gives ride profiles, lengths, elevation, descriptions, tips, and reports.

High Big Cottonwood Backcountry

anywhere. In winter, too, it is one of the Wasatch's most popular Nordic ski-ing spots, with 20 km worth of groomed trails at this facility. In the summer hiking, biking, disc golf, scenic lift rides, and dining are all options. But of course, summer arrives a bit tardy in the high Wasatch, so don't expect to be mountain biking up here in May; with a base elevation of 7,988 feet, it takes a little while before the slopes are clear of snow.

Brighton, founded in 1936, is the uppermost resort in the canyon and is favored among snowboarders and skiers that like to jib in its four ter-rain parks. Outside of these jumps, rails, and pipes, Brighton has 1,745 ver-tical feet over 1,050 acres and the usual eastern Wasatch 500 inches of annual snow—that's almost 42 feet of snowfall in an average year. Open from 9–9 in the height of ski season, night skiing is another of its special assets. In the summer, Brighton welcomes the public to visit for low-key dining, hiking, and fishing. Silver Lake, just a 1-mile walk from the base area, is stocked with trout and may be fished by anyone with a proper Utah conservation license.

To extend this drive all the way east and into **Park City,** backtrack just a few hundred yards from Brighton, and make a right turn to follow Big Cottonwood Canyon Road/UT 190 as it makes switchbacks farther up the sides of the canyon and toward **Guardsman Pass** (closed in winter), the top of which is 16.9 miles from the mouth of the canyon. In the short distance from the turnoff (just more than three miles), the road climbs nearly another 1,000 feet to an elevation of 9,700 feet, changing from pavement to dirt as it zig-zags with grades as steep as 15 percent.

To return to Salt Lake City, either do an about-face at the top of Guardsman, or extend the trip by heading down into Park City. From the top of the pass, the road (now UT 152) winds down through forests, fields, and the chairlifts of **Deer Valley Resort,** T-ing into UT 224. Take a left on this, turning north, and end up in **Park City** (and if you drive even farther north, I-80). Alternatively, head south on UT 224 into **Midway** (the home of **Soldier Hollow Resort,** location of the 2002 Winter Olympic Nordic skiing events) and then, optionally, east into **Heber City.** South of Heber and Midway, US 189 takes a southwesterly direction into **Provo Canyon.** Taking this toward Provo will land you right in the center of the Provo Canyon Scenic Drive (part of chapter 3 in this book). Following this road to the west leads through **Provo Canyon** and into Provo/Orem; turning north on the Alpine Highway/UT 92 toward **Mt. Timpanogos** finishes chapter 3's drive by winding along high, European Alps-esque mountainside and down into **American Fork Canyon.**

IN THE AREA

Attractions and Recreation

Brighton Resort, 12601 Big Cottonwood Canyon Road/UT 190, Brighton 84121. Call 801-532-4731 or 1-800-873-5512; canyons road report: 801-536-5778.

Solitude Resort, 12000 Big Cottonwood Canyon Road/UT 190, Solitude 84121. Winter and summer resort with hiking, mountain biking, disc golf, dining, and lodging available in summer. Call 801-534-1400; canyons road report: 801-536-5778. Web site: www.skisolitude.com.

Dining

Porcupine Pub, 3698 East Fort Union Boulevard/7200 South/UT 190, Salt Lake City 84121. A moderately priced American restaurant and beer pub. Great "halfy hour" and other specials. Family-friendly. Call 801-942-5555. Web site: www.porucpinepub.com.

Other Contacts

Saltlakecycling.com. A highly informative, Salt Lake City–specific Web site with route descriptions and profiles.

Wasatch Wildflower Festival. Taking place for three days, one of which is held at each Brighton, Solitude, and Alta (in Little Cottonwood Canyon) in the middle weekend of July. Call the Alta Visitors Bureau at 1-888-258-2840. Web sites: www.discoveralta.com and www.wasatchwild flowerfestival.org.

Summit of Alpine Loop

CHAPTER

3

Alpine Loop to American Fork Canyon

Estimated length: 26 miles
Estimated time: 2 hours with some sightseeing

Getting There: This route travels south to north, from the Provo area up to the lesser-known Lehi and Highland area by way of Provo Canyon and American Fork Canyon. The starting point for this trip sits in the north-eastern corner of the Provo/Orem continuum, at the intersection of Orem's 800 North Street (UT 52) and University Avenue (US 189). From here, follow US 189 north out of town. Alternatively, to drive the route from north to south, depart from I-15 at exit 284 for Highland/Alpine, and head east on UT 92. This busts directly through sprawling farmland-turned-new-build territory, making a beeline dead east to American Fork Canyon. As of 2010, the cost to enter American Fork Canyon was $6, payable to the Uinta National Forest; annual national parks passes are accepted for this.

Highlights: Popular for both motorists and fit cyclists alike, this mountain route climbs steeply out of the expansive Wasatch Front metropolis and high into this mountain range itself. Beginning at about 4,830-feet elevation, it climbs roughly 3,200 feet to reach an altitude of 8,060 feet. The actual "scenic" business lasts just more than 30 miles, with each end of the scenic drive within about 10 minutes of I-15. This road offers a virtual trip abroad: tight, Tour-de-France-style switchbacks, narrow, looping roads, craggy peaks, and canyon-bottom, riverside driving.

Before departing from town, take a look around and check out what Provo and Orem have to offer. Around the world, Salt Lake City holds the reputation for being the Mormon capital of the world. Though in many ways it deserves this title, Provo also serves many important roles in modern Later-day Saints Church function—particularly in education. Here Brigham Young University and the Missionary Training Center (as well as other institutions) attract a surprisingly diverse collection of Mormons who come from around the world to study and settle here.

Utah Lake, the largest freshwater lake in the state, has about 150 square miles of surface area. Just south of the famous Great Salt Lake, and hardly any higher, this is the last pooling place for Great Basin runoff before the Jordan River flows terminally into the Great Salt Lake. There, water's only escape is evaporation, leaving behind its saline content, and making the Great Salt Lake roughly 10 times as salty as the ocean.

To the east of Utah Lake and Utah Valley, looms the southern portion of the Wasatch Mountains—more than 7,000 vertical feet higher. Mt. Timpanogos, the most famous and visually impressive peak in the range peaks at 11,750 feet above sea level; Utah Lake's surface touches just around 4,500 feet. Mt. Nebo, the southernmost mountain in the Wasatch Range, stands taller than any of its sister peaks at 11,928 feet.

**WINTER ROAD CLOSURE...
EXTENDED!**

Before embarking on this route any time from October through May, be sure to check with **Utah Department of Transportation** (801-965-4000) to be sure the road is in fact open. The lofty elevation of this mountain pass and its more-than 500 inches of annual snowfall close the road each winter—sometimes as early as October, and usually lasting into May or June. Naturally, the time frame is different every year, so be sure to check online for up-to-date information.

Though tall, Nebo is also rather hard to spot; not a distinct, pointy peak, this mountain is recognizable as double peaks on the southern of a long, clean ridgeline.

Beneath the distinctly angular peaks of the Southern Wasatch, narrow canyons drain the massive snow fields that accumulate in winter—as the western side of the Wasatch receives an average of more than 500 inches

Mt. Timpanogos from Alpine Loop Byway

of snow annually. Many mountain streams collect beneath the mountainside, eventually flowing into the valley and adding to the abundant fresh water of Utah Lake.

As Utah Lake contains a vast supply of fresh, mountain water and the climate of Utah Valley behaves rather hospitably, this area has long been inhabited by humans—first lived in by the Timpanogotzis ("Fish Eaters"). Unlike most tribes in North America's interior, who were forever chasing good climes, these fish eaters lived a stable life in one location. The weather neither froze, nor baked, them; the lake fed and watered them.

Their first encounter with Europeans was likely with the famed Franciscan explorer-priests Escalante and Dominguez in September 1776. However, approaching winter chased the explorers out of Utah and back toward

Mexico—they stayed but briefly. Aside from trappers, traders, outlaws, and other transient Europeans, this valley would remain left alone and undeveloped for several decades.

July 24, 1848, would become one of the most significant days in history for Utah and the Intermountain West. After several months of rugged cross-country travel, Brigham Young and his Latter-day Saints followers arrived in the Salt Lake Valley. Being of peculiar beliefs and practices, the Mormons had always endured persecution and societal rejection since their church's founding—especially in the church's pre-1890 days when polygamy was an acknowledged practice. Banished from city after city and state after state in just a matter of decades, the LDS deliberately sought a deserted and utterly isolated place to establish their own society—and even nation—away from other religions and politics. As Utah was utterly isolated—far from any coasts, and also quite far from the Oregon Trail or any major throughway—it would be the perfect location.

Other would-be European residents of the Salt Lake Valley had little interest in it for the exact same reasons that made it perfect for the Mormons—much too isolated and arid, and a challenging environment to coax into sustaining a large population. However, the industrious values and pious nature of these Later-day Saints gave them the perfect credentials for the job. Organized and hungry for a home, they set about voraciously laying down roots. Salt Lake City was pre-planned and under construction by the fall of 1848—just a few months after the historical arrival.

Fields were planted, and plans set in motion. Encouraged by news of success, more LDS pioneers began to arrive by great handcart parades, and church leaders ordered more satellite communities be founded. Colonizing efforts began and pilot groups went out, establishing other towns across Utah in the name of building a Mormon empire.

Provo was established in March 1849. Though at first a small village, it quickly grew in size and importance. The completion of the first transcontinental railroad at Promontory Point, just north of Salt Lake City, would leave many rail workers retired in the Salt Lake Valley, and allow for much easier arrival of non-Mormons to this open, promising landscape. Never being one for "heathens," many Mormons slipped south out of Salt Lake and focused more on the settlement of nearby Provo. While Salt Lake City will always remain the historic capitol of Mormonism, today Provo seems in many ways to be a more spiritually pure, cultural center for the Saints.

In modern times, Provo is the training ground for blossoming Mormons. On its soils sit two of the church's most important educational centers, **Brigham Young University** and the **Missionary Training Center** (actually part of BYU). BYU and the MTC remain especially connected to the church, as both are funded by it and were established expressly for church members. Though Orem's **Utah Valley University** is technically not a Mormon institution, many thousands of the area's youth receive a degree from it every year. Together these institutions occupy roughly 70,000 students and faculty.

Though technically two separate townships, Orem and Provo operate nearly as one, and flow into each other seamlessly, sharing many of the same streets and community facilities. Anyone requiring food will notice that the area is teeming with national, mid-level chain restaurants. For a locally flavored bite, head to the intersection of **University Avenue** and **Center Street;** park the car and walk in any direction to find dozens of small eateries featuring almost as many cuisines. Given the general monoculture ambience of Utah Valley, it may come as a surprise that this small intersection features restaurants representing the cuisine of countries from Korea to India, most of which are run by first-generation immigrants. Why so? As the Mormon Church spreads around the world, international members have been inspired to live in Utah, quite often specifically for an education at BYU or to be trained at the MTC.

Lodging in Utah Valley is almost exclusively of the budget-to-average national brand variety—dependable yet unadorned. The one major exception to this rule exists outside of Provo at the **Sundance Resort** (described later in this chapter). Sleepy visitors in the valley can choose between any imaginable chain, from a Hampton Inn to a Best Western and the Marriott. As with most of the valley's restaurants, these have been clustered around University Avenue, downtown, and near I-15.

Head toward Provo Canyon and out of the city from Orem's 800 North Street (UT 52) and University Avenue (US 189). Those wanting to check out the canyon as a pedestrian, rollerblader, or cyclist can pick up the Provo Parkway here or at any of the numerous access points between Utah Lake and its terminus at Nunn's Park. (For more information, visit www.utah mountainbiking.com and search for this ride.) This paved, 14-mile trail heads up and into Provo Canyon, providing a peaceful way to get some exercise and enjoy nature away from US 189. Tracing along the Provo River, this pathway climbs into the canyon, but it has a very tame grade and is extremely popular among casual cyclists, joggers, families, and strollers. Be aware that in town the path often plunges underneath streets via narrow tunnels, and careless riders with too much speed can cause a serious collision with others.

Drivers: point the wheels north/northeast and head out from Orem on US 189. This range is overseen by the Uinta-Wasatch-Cache National Forest. Leaving Utah Valley, this pavement ribbon begins to ascend into **Provo Canyon,** which is drained by the **Provo River.** Though the Wasatch Range isn't necessarily famous for fly fishing, Provo River offers some of the best trout fishing in the area. Its entire course consists of roughly 70 miles from the Uinta Mountains' **Washington Lake** at 10,450 feet elevation, down to Utah Lake. In the lower and middle sections of this river swim brown, cutthroat, and rainbow trout—roughly 5,000 fish per river mile. For a guided trip, equipment retail or rental, or just online fishing forecasts, check out **Four Seasons Fly Fishers,** located in Heber City or call the **Park City Fly Shop** in…Park City.

As the US 189 climbs into the canyon, the first stop along the way is

---

CYCLISTS' DREAM

For hardcore road cyclists, the Alpine Loop is one of Utah's finest mountain rides. Cyclists can follow more or less the exact same path as this chapter's driving route, looping easily back south through town at the "end" of this scenic drive.

To begin the ride, cyclists can pick up the Provo Parkway, a paved pedestrian and bike path, at an access point right at the mouth of Provo Canyon on 800 North. Head east into the canyon, and rejoin the main road and scenic route in the **Bridal Veil Falls** area.

As the Alpine Loop climbs high onto the shoulders of Mount Timpanogos, the way becomes quite narrow and must be shared with vehicles. Because of the steepness and winding nature of the road, most drivers should be maneuvering attentively—but that should never be taken for granted and bikers should be extremely cautious.

During spring a magic window of good weather sometimes avails itself in May. This is a time when the road is still closed to vehicular through-traffic, but is clear of snow. During this period, and before the true heat of Utah summer kicks in, this can be one of the most enjoyable paved bike rides in the state, utterly scenic and free of traffic.

Kids Playing beneath Bridal Veil Falls

Canyon Glen Park—just one in a series of many city park-type alcoves with tables and green space in this canyon. This particular locale's bonus for picnickers is its distinctly smaller crowds than those in the Bridal Veil Falls vicinity. From this park you can also access the **Great Western Trail.** This enormous, often very rugged path leads all the way from Mexico to Canada through the Western United States, crossing over five states: Arizona, Utah, Wyoming, Idaho, and Montana. It adds up to 4,455 miles, yet is still wild, rough, and sometimes crudely patchworked—much less developed than its more famous cousin to the east, the Appalachian Trail. Sixteen hundred miles of the GWT pass through Utah, incorporating many of the state's most popular and scenic trails near Bear Lake, Brighton Ski Resort, Mount Timpanogos, Fish Lake, Bryce Canyon National Park, and Grand Staircase-Escalante National Monument. Still a work in development, check out the trail's Web site for maps, blog links, and images.

Three miles into the canyon is the location of one of its most famous attractions, **Bridal Veil Falls.** Off this two-tiered, cataract falls, a ribbon of water cascades steeply down vertical limestone cliff bands for a total of 607 vertical feet. Clearly visible off to the south side of the road, these can also be seen by taking the short, paved hiking trail that ventures yet closer from the signed pull-off for Nunn's Park/Bridal Veil Falls. To hike the trail, turn right (and away from Nunn's Park) from this exit; hike up canyon on the most obvious broad, paved path. The falls are about half a mile away.

During winter, keep an eye out for ice climbers on this flow. Though there are numerous ice climbing areas in the canyon, this is one of the most popular. This famous ice climbing route, called Stairway to Heaven, is so named because of its terraced nature and is one of the more famous and tall water ice climbs in the United States. Many climbers stick to the first pitch of the route called the Apron, but climbers wishing to continue can go on for multiple pitches above that.

Nunn's Park, just across the highway and under the bridge from Bridal Veil Falls, makes a good base camp for those checking out the falls. Like a typical city park, this has shade trees and green grass. Here there are more than 12 campsites (for $15 per night), pit toilets, and different pavilions for hire. All camping is handed out first come, first served; pavilions must be reserved and paid for ahead of time. The park's namesake, L. L. Nunn, was an early hydroelectric power thinker, who operated the first 44,000 volt water-powered plant in America here—back in 1897. Converting the potential and kinetic energy of the Provo River into electricity, he provided power for mines in the Mercur, Utah area. There are still tidbits left over from the hydropower plant days—as well as a plaque to commemorate its existence.

Next up is **Vivian Park,** just fewer than 6 miles into the canyon (and about 2.5 up from Bridal Veil Falls). Historically, this area has been an attraction for travelers and recreationalists of various wants and needs since 1888. Originally, Billy's Place was a place for Provo Canyon travelers to dine and recover. Around 1900, its role switched to that of a canyon retreat, complete with restaurants, guest cabins, and live music. When the world started getting smaller, people could travel farther away to bigger resorts, such as Sundance Resort (later in this chapter), and this retreat became defunct. Today a tiny village of miniature-looking homes is tucked into the forest here. Nearest US 189 is the public recreation area with picnic tables, pit toilets, and greenery. There is also a fishing area (for children 12 and under only). From the west end of the park (departing from its central parking lot) is the **Provo River Parkway** (as described earlier in

---

BRIDAL VEIL FALLS TRAM

From 1967 through 1996, a six-passenger aerial tram shuttled guests from Bridal Veil Falls parking area to the Eagle's Nest Lodge, atop the falls. The lodge wore many hats throughout its lifetime, and the tram was the only reasonable means to reach it. But both the tram and lodge were set up in a difficult area with fickle weather and business. The tram was claimed by locals to be the world's steepest, but hard evidence is lacking to this end.

The lodge itself changed ownership many times, and during its tenure from roughly 1961 to 1996, it functioned sometimes as a restaurant, sometimes as a special events venue. The nature of its location—precipitously perched on a mountainside—left it and the tram quite exposed to the elements, particularly avalancheq. When one such slide finally destroyed the tram in early 2006, the restaurant, whose only access had been via the tram, closed "temporarily." Unable to raise funds to restore and reopen, the owners finally gave up trying to reopen in 2007 when the Forest Service required a mandatory $100,000 environmental study to be conducted before it would consider allowing a reopening.

The final blow came in July of that year, a wildfire burned 240 acres in the vicinity, including the restaurant itself, as well as four of the five tram cables; the other was later snipped and removed. Today the site remains vacant, with few good solutions in the air for future use of the land.

---

this chapter). This 14-mile paved pathway leads back down the canyon, through town, and to Utah Lake.

Just about 7 miles into the canyon, and still barely above the 5,200 feet of elevation, the route departs from Provo Canyon, turning left (northwest) onto the Alpine Loop (UT 92) and begins to seriously climb. Just 2 miles after this turnoff comes the **Sundance Resort,** nearly 2,000 feet higher than the Alpine Loop–US 189 junction just a few miles back. For most, the name "Sundance" signifies more than just a ski resort. Sharing the same name and founder with the Sundance Film Festival, this Robert Redford ski area is one of the smallest and most beautiful in Utah. Much more of a retreat than a tram-driving, crowd-pushing resort, the restaurants and arts culture are the only features that compete with its natural beauty.

Robert Redford, who famously played the Sundance Kid in *Butch Cassidy and the Sundance Kid,* became enamored of Utah during his many shoots in the area. He purchased a fledgling ski area on this site in 1969 and transformed it into the modern-day resort. Today, it's a peaceful stop in any season if for lunch or even a multi-day stay. Everything at Sundance—from the cuisine to the architecture—draws from and highlights the natural environment.

The dining at Sundance is some of the best in the area, and people regularly drive from Provo and beyond just to enjoy dinner here. Those needing sustenance may choose from many options. Two restaurants, the Tree Room and the Foundry Grill, both present an option to sit down and enjoy local game and vegetables, as well as seafood, fresh fish, and more. Hearty, seasonal cuisine is chefed up with fine skill at both places, with the **Tree Room** being the much more formal of the two. This four-diamond, Wine Spectator

SUNDANCE FILM FESTIVAL

Though you won't find the festival here at the Sundance Resort, the fact that the two share the same name is not a coincidence. Like the resort, the now world-famous independent film festival was also the brainchild of actor Robert Redford—a man that has led a life using his stardom to promote the arts and philanthropy. Established in 1981, this is the most influential independent film festival in the world to date, taking place primarily in Park City, but also flowing over into Salt Lake City and beyond. Sundance fills the local theaters in January, and its busyness gridlocks the small mountain town. Tickets are nearly impossible to get last-minute, but with advanced planning or a steel will, they can be acquired. Good luck parking, though!

Award–winning affair serves modern American and Continental cuisine with a Western flair and heavy use of local, organic, and generally top-notch ingredients.

The **Foundry Grill** provides a meal to those wearing jeans, towing children, or simply desiring a less formal or pricey experience. Also serving regional American cuisine, this menu is nothing to pooh-pooh. The offerings span the range from salad, pizza, and pasta, to halibut and filet mignon. But this is no iceberg lettuce salad bar or greasy pizza; the salads are lush, green, and creative, and the pizza is wood-fired.

For those in a rush or wanting to move on, **The Deli** is a Sundance-flavored general store–type affair, offering everything from gift baskets and wine to chilled soda, specialty sauces, dressings, and preserves, fresh baked goods, and other snacks. Those not wishing to drive any farther for the day should consider unwinding with a beverage. In the resort center, the **Owl Bar,** serves libations nightly across a historic Victorian rosewood bar that was actually used back in the 1890s in Thermopolis Wyoming—a watering hole that Butch Cassidy and the Wild Bunch often visited.

While at the resort, there are numerous activities to check out any time of the year. By summer the resort offers fly fishing, mountain biking, hiking, cross country horseback riding, and moonlit lift rides. By winter, snowboarding, Nordic and downhill skiing, obviously draw most of the visitors. But year-round, the resort sports a spa and hosts various nature programs like night owling and nighttime snowshoeing, as well as children's programs. The Sundance Spa has gained national notoriety, and has been featured in a number of magazines such as *New York, Travel & Leisure,* and *Spa.*

The **Sundance Art Shack** is another year-round affair completely atypical for a ski resort. Here work artists in residence, whose art can be seen in exhibits around Sundance. These artists also oversee and instruct daily, two-hour workshops designed for guests. Check out the glassblower studio and take a class there if time allows. The gallery and glassblower studio are both located right in the Art Shack, and clinics are offered multiple times daily.

After leaving Sundance, prepare for ever steepening and narrowing roadways as the pavement now begins to ascend into the high shoulders and cirques of the tall Wasatch and Mount Timpanogos. The steepness of the terrain forces the road to narrow severely and fold into switchbacks; at points, the vistas seem nearer to those of Europe's Alps than central Utah.

AMERICAN FORK ROCK CLIMBING

Look around on this leg of the trip and notice the many dozen limestone cliffs rising at all heights throughout the canyon. Though many of these cliffs appear rotten and crumbly, a few dozen of them actually consist of relatively compact, sturdy rock. The faces and caves of these "better" cliffs have been a destination for rock climbing for decades.

Technical rock climbing, which arose out of mountaineering, began in a much different style than that done on these gray rocks. As rock climbing moved away from mountaineering and high summits, a new sport arose on sheer rock faces, pushing physical and technical difficulty. First following cracks, these routes stuck to the natural "lines" in rock and could be protected with removable gear. Yosemite is one of the famous birthplaces for rock climbing in North America during the 1960s and 1970s.

However, during the 1990s, a new type of climbing became popular that didn't require a crack for placing protection. Sport climbing, as it's called, implemented the use of bolts drilled into rock. Climbers, now able to clip their rope into these bolts with carabiners, could now climb faces of rock—totally independent of removable gear.

Sport climbing, by nature, was much faster and more accessible than traditional climbing. It required much less gear and emphasized placement of protection much less. Rather than spending time finding the right cam for a crack, a climber could simply clip a pre-placed bolt and move on. Many "traditional" climbers find this new sport perverse, and some consider it invalid.

Rock Climber at the Division Wall in American Fork Canyon

With the new ease and speed of protection, however, climbers could now focus on harder moves and routes. "Whippers," where a climber falls off a route, became the norm as climbers could now "work" routes and moves, until they gained the fitness, skill, and memory necessary to "send" them, or do them in one push from the ground.

Since sport climbing's inception in the '90s, it has revolutionized the sport, pushing climbers beyond what would they could do with traditional gear. Though many extremely difficult routes have gone up on traditional gear, it is thanks to sport climbing that these first ascentionists had the strength and skill to apply to traditional climbing.

Though today many sport climbing areas have been developed around the country, this canyon is special. American Fork Canyon was once one of the sport's pioneering epicenters in the United States and is therefore considered a historically significant location for climbers around the world. So even though the routes are polished, climbing snobs should appreciate it for what it's worth and imagine bolting some of these lines back in the 1990s!

Popular walls today include the Hell Cave, home of some of the canyon's most difficult routes, the Division Wall, with many 5.10s–5.12s, and the Escape Buttress, a south-facing 5.10 haven. For a vastly more thorough guide, check out Climber's Guide to American Fork and Rock Canyons, by Stuart and Brett Ruckman.

From here, it's hard to believe that just 60 miles west, as the crow flies, lay the salt flats of the Great Salt Lake Desert and the Dugway Proving Ground.

Just 6 miles beyond the Sundance Resort, the Alpine Loop Road tops out. At elevation 8,030 feet, the Alpine Loop is draped over the high shoulders of Mount Timpanogos, a broad fortress of a peak. Take a deep breath (of thin air!) and begin to descend. Keep speed in check and watch for oncoming traffic and bikers—this is one of the more popular bike rides in Utah.

Six miles of truly stunning descent brings the road through Pyrenees-style switchbacks and down almost 2,000 vertical feet (to an elevation of 6,070 feet)—a drop symmetrical to the climb on the southern side of the pass. At the bottom of this steep decline, this road meets up with **American Fork Canyon.** To enter the canyon, a day fee of $6 per vehicle (as of 2010) must be paid at the **Uinta National Forest** toll booth; national parks passes are accepted.

American Fork Canyon, like all the major Wasatch drainages, runs east-west and cuts deeply into the tall range, carving severely steep, narrow slots into the already impressive, craggy Wasatch Range. Though not many people find themselves readily occupied in the canyon—no overlooks, no restaurants, no gift shops—it nevertheless is a beautiful canyon in and of itself, offering multiple Forest Service picnic grounds and campgrounds. Beware here—though it is not overtly dangerous, yearly trail injuries, animal attacks, and weather incidents cause injury and death. Education on basic wildlife behavior and taking precautions can help avoid these situations.

Heading down American Fork Canyon and toward Lehi (the nearest township in the Utah Valley), visitors pass by **Timpanogos Cave** very near the western end of the canyon, near its mouth.

Timpanogos Cave National Monument represents one of the largest, most publicly accessible caves in the region. Though the approach to the actual cave does require physical effort, the hike can be called "reasonable" by most standards. To reach the cave, all visitors must follow a 1.5-mile, paved path that gains almost 1,200 feet of elevation from the parking lot to the cave's mouth. Though the path becomes quite steep in some places, it remains broad and smooth. Climbing out of the canyon floor and onto its southern slopes gives perspective and showcases the steepness of the v-shaped mountainsides.

The cave typically opens May through October. During this time the park operates tours through three major chambers: Timpanogos Cave, Middle Cave, and Hansen Cave. To enter, all visitors must be part of a guid-

---

BEAR ATTACK: A WORD OF CAUTION

Though American Fork Canyon is overwhelmingly safe, it doesn't hurt to remember that it is also home to wild animals. In the early summer of 2007, a family camped at the Timpooneke Campground to celebrate Father's Day and christen the new tent that Dad had received as a gift. That night a black bear made its way into the campground, likely lured by the scent of food in the campsite. The animal mauled a sleeping 11-year-old boy, dragging him from the tent and killing him.

At the time of the incident, this attack stood as the first lethal bear mauling in the state's written history. This type of attack can largely be avoided by keeping the campsite clear of enticing scents—by tying food high up in a distant tree or keeping it locked in the car, well away from the campsite—and carrying bear spray at all times. Regardless of precautions and existing warnings, wildlife is always a concern and should never be fed—for the sake of the animal and of the people.

---

ed trip. These cost $7 per adult ($5 for junior up to 15 and $3 for children under 5; free for infants 2 and under) and can be reserved by calling the park. During the summer, reservations are highly recommended.

The first of the caves to be discovered was Hansen Cave, found by cougar-tracking Martin Hansen, all the way back in 1887. In the intervening time, before coming under government protection, much of Hansen Cave was vandalized and its features destroyed. Timpanogos and Middle caves remained just a rumor until 35 years later, when they were each discovered by separate parties. Because of their delayed discoveries, these two caves' interiors were better protected and remain well preserved today. The interior of the limestone complex is densely featured with speleothems, including stalactites and stalagmites, cave bacon, popcorn, flowstone, and rock straws called helictites.

Leaving the Timpanogos Cave and heading west, the ride is all but done. Soon enough the canyon opens up, quite abruptly, dumping traffic into Utah Valley and the suburbs of Highland. To return to the interstate, stay on US 92 which travels due west, intersecting with I-15 in about 8 miles—no turning necessary. If biking, head south/southeast on any number of roads to return to the car. The main roads heading back toward Provo Canyon are UT 146 (Canyon Road), later 100 East Street, US 89 (State Street) and finally UT 52 (800 North Street, heading east). The total road bike loop is almost exactly 70 miles.

IN THE AREA

Accommodations

Sundance Resort (See "Attractions and Recreation" below)

Attractions and Recreation

Four Seasons Fly Fishers, 44 West 100 South, Heber City. Retail and rental shop, guide services. Call 435-657-2010. Web site: www.utahflyfish .com.

Nunn's Park, Utah Country Park Department. For pavilion reservations or general questions, call 801-851-8640. Web site: www.utahcountyonline .org.

Sundance Art Shack, Sundance Resort Center (see Sundance Resort). Gallery open Saturday and Sunday only; Art Shack open daily, check Web site for workshop schedule and to make reservations. Call 801-223-4535.

Sundance Resort, 8841 North Alpine Loop Road, Sundance. Call 1-866-259-7468 (or **Tree Room Restaurant**, 801-223-4200, and **Foundry Grill**, 801-223-4220). Web site: www.sundanceresort.com.

Timpanogos Cave National Monument, R.R. 3 Box 200, American Fork. Call 801-756-5239 (Headquarters) or 801-756-5661 (summer only). For tour reservations (recommended) call 801-756-5238. Web site: www.nps.gov/tica.

Vivian Park, 5828 North South Fork Road (US 189), Provo Canyon. For pavilion reservations or general questions, call: 801-851-8640. Web site: www.utahcountyonline.org.

Dining

The Deli, Sundance Resort Center (see Sundance Resort). Open daily, morning through evening. Call 801-223-4211. Web site: www.sundance resort.com.

Foundry Grill, Sundance Resort Center (see Sundance Resort). Open for breakfast, lunch, and dinner Monday through Saturday, and brunch on Sunday only; closed between meals. Call 801-223-4220. Web site: www .sundanceresort.com.

Owl Bar, Sundance Resort Center (see Sundance Resort). Open evenings Monday through Friday, and earlier on weekends. Call 801-223-4222. Web site: www.sundanceresort.com.

Tree Room, Sundance Resort Center (see Sundance Resort). Fine, formal dining; American and continental cuisine. Open for dinner only Tuesday through Saturday, closed Sunday. Reservations encouraged. Call 801-223-4200. Web site: www.sundanceresort.com.

Other Contacts

Brigham Young University, 150 East Bulldog Boulevard, Provo. Call 801-422-4636. Web site: www.byu.edu.

Great Western Trail. Trail system spanning the Western United States from Mexico to Canada, passing Arizona, Utah, Wyoming, Idaho, and Montana on the way. Web site: www.gwt.org.

Missionary Training Center, 2005 North 900 East, Provo. Call 801-422-2602. Web site: www.mtc.byu.edu.

Park City Fly Shop, 2065 Sidewinder Drive, Prospector Square, Park City. Gear retail and rental, guide services. Call 435-645-8382 or 1-800-324-6778. Web site: www.pcflyshop.com.

Sundance Film Festival, Sundance Institute, 1825 Three Kings Drive, Park City. Call 435-658-3456. Web site: www.sundance.org.

Utah Department of Transportation. Responsible for road maintenance, issuance of reports. Call 801-965-4000. Web site: www.udot.utah .gov.

Utah Valley University, 800 West University Parkway, Orem. Call 801-863-7149. Web site: www.uvu.edu.

Nearing Logan from the South on US 89/91

CHAPTER

4

Logan Canyon to Bear Lake

Estimated length: 39 miles from Logan to Bear Lake
Estimated time: Half a day (with return trip)

Getting There: Logan is one of just a handful of medium-sized towns in Utah not attached to an interstate or an urban conglomerate, and yet not deriving its main industry from tourism. As it is separated from I-15 by a mountain ridge and about 20 miles (as the crow flies), there is no single way to get there. If coming from the south, the fastest way is via US 89/91, departing from I-15 in Brigham City. This major, oft-divided highway heads northeast, climbing up high through mountain passes before dipping back down again into the Cache Valley. Coming from the north, access can either be had by taking US 91 south from Idaho and directly into Logan, or by taking UT 30 east from I-15 at exit 385.

Highlights: Along the way, the road climbs up and out of **Logan,** through **Logan Canyon** and along the **Logan River.** Ascending more than 3,000 vertical feet to a crest of 7799 feet, it then descends into the **Bear Lake Valley** toward **Bear Lake,** which spans the Utah-Idaho border and is the second largest fresh water lake in Utah (behind Utah Lake). The more the road climbs, the fewer and farther between the city-park-like picnic grounds, and the more the landscape transforms from typical Utah mountainlands to Montana- or Wyoming-like hills. Narrow twists and cliffs at the bottom of the canyon give way to a broader landscape at the top with rolling, grassy slopes.

Logan is one of Utah's neatest, most genuine mid-sized towns. Not a dot on a map, a tourist town, or a sprawling suburb, this university town has roughly 49,000 residents and a beautiful post in the **Cache Valley** next to **Logan Canyon,** which cuts deeply into the **Bear River Range.** At the colder end of Utah's diverse climate spectrum, this has seasons more closely resembling those in Wyoming than other towns in Utah.

The Top of Utah Marathon, one of the area's most famous events, is a 26.2-mile suffer fest (with shorter versions) that begins in **Blacksmith Fork Canyon** at 6:55 AM on the third Saturday in September each year. Starting at **Hardware Ranch,** a refuge for moose, elk, and deer during the Bear River Mountains' harsh winter months, it runs down this canyon along UT 102, and finally into town. This has become quite popular among runners for its supreme beauty and, possibly, because it's a gradual downhill course.

Hungry people in Logan need look no farther than **Main Street.** This north-south running strip of pavement is littered with restaurants of every genre and price range. For a statewide favorite, visit Logan's **Indian Oven.** Opened by a native of India—also a restaurateur since 1989—this casual but handsome and newly remodeled restaurant offers an extensive menu of authentic Indian cuisine with ample vegetarian offerings. Open for lunch and dinner only, the restaurant unfortunately closes its doors on Sundays.

Le Nonne is Logan's favorite dispensary of Italian cuisine, serving traditional, high-end fare in an upscale, remodeled, historic home. This place is yet another favorite in the area—and actually in the state—and is always one of the top-ranked restaurants in northern Utah. Again, this restaurant closes on Sundays, so plan ahead if planning to visit. Primarily a dinner establishment, lunch is served on Thursday and Friday. Make a reservation!

If you need a place to crash, Main Street is again the place to go. Many national hotel and motel chains have assembled along this street. **University Inn** represents Logan's locally run, hotel-style quarters. Owned by **Utah State University,** this classic and stately hotel offers a range of rooms, from "guest room" to deluxe suite. Right on campus this can be a relatively serene location in-town, but not on the weekend. The **Beaver Creek**

Downtown Logan

Lodge is one of the more unique deluxe hotels in the area, but it is in Logan Canyon (27 miles east of town; described later in this chapter).

While sleeping or dining on Main Street, it doesn't hurt to stop in at the **Cache Valley Visitors Bureau.** This group distributes a wealth of information (some free, some visitors must pay for) pertaining to Logan, Logan Canyon, Cache Valley, Bear Lake, and Utah in general. Even if just driving through and never leaving the car outside of town, illustrated maps can shed some light on the significance of the views.

US 89 enters **Logan Canyon** by heading east from the city center as 400 North Street. This canyon, carved by the **Logan River** (ah *ha!*), is also cooled subtly by it during summer, bringing temperatures here to a more humane level than in the sun-baked valleys. About 39 miles long from Logan to **Garden City,** US 89 parallels the river, climbing high through different landscapes, and ending in the Bear Lake Valley, near the shores of the strikingly blue-green Bear Lake.

Before plunging into the Bear River Range Canyon, take a glance back across the Cache Valley and notice the very obvious horizontal terraces ringing the valley. During the most recent ice ages that receded as recently as 12,000 years ago, this valley was submerged under hundreds of feet of Lake Bonneville's waters, just as the Salt Lake Valley. The cool and moist climate of that time caused the **Great Basin** to fill with a lake that spanned from **Cedar City** to **Red Rock Pass** (near the Utah-Idaho border and I-15)—350 miles in length, and almost half as wide. Modern day Logan would sit under more than 600 feet of water. During these times, the fauna was vastly different than today, with musk oxen, wooly mammoths, bison, and camels cruising the shoreline.

Those with any questions about the vast expanses of the **Uinta-Wasatch-Cache National Forest** (which is where the route is about to enter) should stop by the **Logan Ranger District Visitors Center** on the way. Situated cleverly at the mouth of the canyon, right on the roadside and well signed, it is open in all seasons (but closed on weekends!). Here visitors can pick up maps, ask about camping, get advice on recreation, or purchase keepsakes and postcards.

The first several potential stops happen in quick succession and right after entering the canyon. The first in one of a series of three low-head dams (cleverly named "**First Dam**") pooling the Logan River's waters. Around this mini reservoir sits a city park with picnickers, casual fishermen, and the like. Built up with bathrooms, fishing docks, and such, the park is well manicured and user friendly.

Those wanting to get stoked about the flora and fauna in Logan

UTAH STATE UNIVERSITY

Unlike in many isolated towns in Utah with a substantial economy, Logan doesn't base its commerce on mountain biking and outdoor recreation—even though there's a lot of that to be done around here. Instead, one of the main attractions to this area is **Utah State University,** which does its job bringing in people from across the state, country, and even the world.

Founded in 1888, this university has spent more than 120 years growing from a tiny agricultural college, to a national authority in ecology and other environmental studies. In total, the enrollment of the institution is 25,000 undergrad and graduate students (with 20 percent of these being students at satellite study sites across Utah). Its graduate school in these fields is quite prestigious, and for a decade has been ranked in the top 2 percent in the United States and top 500 universities in the world.

Limestone Spires in Lower Logan Canyon

Canyon should move right along to the **Stokes Nature Center,** just 0.6 miles upriver of the canyon's mouth. From fish to birds, this organization educates any curious person on Logan Canyon's environment—and it can enrich an otherwise uninformed drive. A largely program-based operation, this center offers many educational events to the general community, in addition to schools and groups. Check online for the current schedule. Park on the south side of the road and walk another tenth of a mile up canyon to pick up a short trail across US 89. This trail leads to the center after just another few hundred yards of walking. From here, too, the **River Trail** can be picked up. Perfect in length and pitch for a jog or casual walk, this 4.2-

Middle Logan Canyon

mile trail runs parallel to the Logan River and is sheltered from the highway by trees and bushes. Use your newly gained knowledge from the nature center to observe the canyon bottom.

Moving up canyon, the trail, road, and river all pass the **Second Dam** and the **Logan City Power Plant,** in operation since early 1900s—during a time when power bills were assessed based on the number of light bulbs operating in a household. Today this turbine provides the valley with roughly a sixth of its power. This pullout also offers another chance to picnic and use the restroom; but it won't be the last opportunity.

Three miles into the canyon is the first car camping opportunity, **Bridger Campground.** Located in the Rocky Mountains at 5,400 feet, snow closes this sucker down every year, typically between October and May. If arriving during a questionable season, call the Logan Ranger District Visitor Center to double-check.

The **Third** (and final) **Dam** is the next stop, where from another campground, **Spring Hill,** the hiking options start to branch out a bit. On the eastern edge of the campground, the **Riverside Nature Trail** follows the Logan River. Along it, small placards interpret the environs, identifying and naming different plant and animal species. Wetlands near the campground render the area abundant with waterfowl and other critters not extraordinarily common in dry Utah.

The other path leading away from this area is the 3-mile **Crimson Trail** that heads to **Guinavah-Malibu Campground.** Called so for ye olde Brigham Young College. Formerly based in Logan (now Brigham Young University in Provo), its graduating students held ceremonious walks along this trail, wearing the school's then red and gold colors. Along the way, this steep trail passes by a striking band of limestone called the **China Wall.** As limestone is fossilized seabed, it often contains preserved sea creatures. This section of rock happens to contain a lot; take the time to look for and find them. Connecting to the Riverside Nature Trail, this route loops right back to the Spring Hollow campground, after a total of 4 trail miles.

Climbing higher into the canyon, more campsites and picnic areas continue to litter the roadside. A little more than 4.5 miles after US 89 enters the canyon, it passes by **Wind Caves** and **Cave of the Winds Trail.** (These caves are also visible from the DeWitt Picnic area, back down the road about a quarter of a mile.) It may surprise Utahans that sandstone is not the only type of rock that forms arches; here, wind and water have carved a triple arch into limestone. This cave-like formation seems like a gazebo with sky lights. The trail to reach the caves stretches about 2 miles each way and provides a more panoramic, zoomed-out view of the canyon below.

Just about 19 miles into the canyon, its formerly narrow floor broadens in the **Tony Grove** area. Not named for a gentleman called Tony, this portion of the canyon actually became known as such because of group of late 19th-century elite (or "toned") group that regularly vacationed here during the summer. Taking the Tony Grove Canyon Road to the left (northwest), park after just 0.25 miles. Here stands a 1907 log structure that served as the home of the area's first ranger. Now part of the **Tony Grove Memorial Ranger Station Historic District,** this cabin stands in the company of a cluster of newer buildings. Erected in the 1930 for all kinds of forestry use, these served as training and guard stations, as well as nurseries. Still in operation today, the structures have been maintained throughout their life.

Those who wish can continue along US 89 toward Garden City, others can embark on a deeper-yet detour up and to **Tony Lake,** which sits near the perimeter of the **Mount Naomi Wilderness.** This 45,000-acre protected area was set aside as such in 1984 by the United States Congress. A total of 7.5 miles of pavement climb from US 89 to this glacial lake, which sits at an elevation of 8,043 feet. A natural lake by origin, this has also been enlarged by the construction of a dam in 1939—though it's still quite small, at just more than 25 acres. Tony Lake is supplemented with pit toilets, picnic tables, and a few dozen campsites. Fishing is possible here, but the high elevation of the lake serves to kill many fish during the cold winter months.

At the lake, a signed nature trail traipses along the shores, depicting a bit of the area's wildlife and geology. A "backcountry" trailhead here serves as the gateway to a number of different hiker- and horse-friendly trails. Heading to the northeast, a trail leads to **Naomi Peak,** elevation 9,979 feet (gaining roughly 1,900 vertical feet over 4 miles); to the south a trail leads to **White Pine Lake,** passing various high elevation springs and ponds. Because of the general moistness of this area, moose are a common sight every year; beware: though these animals appear goofy and docile, they in fact can be quite fierce. Though they won't bite or eat a person, they will charge, ram, and trample anyone—particularly a person too near, or interfering in any way with, a baby moose (called a "calf"). To park here, all vehicles must pay a few dollars for a day-use fee.

About 3 miles and a few more campground/picnic areas farther along

Broad Top of Logan Canyon

US 89 are **Franklin Basin** and the **Steam Mill Hollow Trail.** Heavy beaver trapping between 1820 and 1840 nearly wiped out the species in this area (not unlike in the rest of the country), but intervening decades have allowed them to repopulate. Therefore curious visitors should keep an eye out for their dams. Made of sticks, these hollow, dome-shaped mounds block water flow and, like small huts, actually provide a home to the beavers. Anyone that hikes long enough may get lucky enough and stumble upon buried, stolen gold. Apparently during the late 1800s, two robbers evading a posse fled into these hills and buried a serious quantity of gold here.

Now 25 miles from the mouth of the canyon on the main road is the turnoff to **Beaver Mountain.** The northernmost ski area in Utah, Beaver Mountain gets about 20 percent less snowfall as the famous Wasatch resorts such as Snowbird and Alta (still more than 400 inches annually), has 1,600 vertical feet of drop, and almost 700 skiable acres. But with an insulating drive from Salt Lake City and Park City, it also has vastly less crowding. The resort, which was first opened by the Seeholzer family in 1932, is still run by the family today. In its early decades, the resort experienced growing pains and changed locations a few times, but it seems to have permanently settled where it is, there since 1949.

During winter, snowmobilers (in addition to snowboarders and skiers) come to Beaver Mountain. In summer, it is possible to hike and mountain bike there, as well as stay in guest accommodations (yurts or heavy-duty, permanent tents) or in its full-service RV park. Check online or call for an event schedule; sometimes the resort even hosts summer concerts. For more lodging, head up the main highway another half a mile to the **Beaver Creek Lodge** (a total of 27 miles up Logan Canyon), which offers more typical guest lodging.

This log structure fits perfectly in this Rocky Mountain forest. The simple, handsome rooms keep in theme with log furniture and quilts—but the modern structure is spacious and offers guest plenty of light. Popular in all seasons, this is a snowmobiler's delight in winter, and a horseback-riding haven during snow-free seasons. Oops! Forgot the snowmobile or horse? Rentals are available at the lodge itself. Optional packages include lodging, equipment (or animals) appropriate for the season, and meals. It may be worth considering hiring a snowmobile, as approximately 300 miles of groomed trails in the vicinity of the lodge make this one of the nation's top 10 "sledding" destinations.

The next major pullout is for **Swan Flat** on the left side of the road. This road system, departing to the north (left), sends dirt spurs up toward the Idaho border. As is obvious from its appearances, this area is a gateway to undeveloped, outdoor recreation. Nearing the highest, broadest part of Logan Canyon, these roads can be snowmobiled or cross-country skied in winter and ATVed, biked, ran, or hiked in the summertime.

A trail departing from the road (about 3.5 miles north of US 89, and almost in Idaho) leads to **Mount Bridger.** Named for the famous mountain man, Jim Bridger, as with many other mountains and towns, from Montana and Idaho, to Wyoming, and Utah (see sidebar, below), this mountain is the tallest in **Rich County.** The trail to it begins as an ATV path, but eventually thins. To reach the actual peak itself, hikers must use their judgment and depart from the trail, following the final ridge to the actual peak. A cairn and summit register marks the spot.

Speaking of Rich County, it should be noted that this is quite a peculiar section of Utah. During the 1990s, the *Salt Lake Tribune* was penning an article about the tallest mountains in each of Utah's 29 counties. At that time, the 9,255-foot-tall Bridger Peak was yet unnamed, and only became so for the sake of the article. Though it is the shortest of all the counties' high peaks in Utah, it is still (here comes some trivia) the highest of all the shortest county high points in the United States. If that makes any sense!

This area is also the home of some strangely cool weather—as well as the spot where Utah's all-time low temperature was recorded. On February 1, 1985, **Peter Sinks** posted a low of -69 degrees F—actually the lowest temperature ever measured in the continental United States. Even in the summer, it is rare that Peter Sinks experiences more than three consecutive days without the thermometer dipping below freezing. Not the only "sink" around; this is part of a wider-spread geological phenomenon (see sidebar, below).

Moving farther along US 89, the road swoops back toward the south.

JIM BRIDGER

In his day, this man traveled a remarkable amount. A well respected leader wherever he went, Bridger spread his name everywhere, especially in the Rocky Mountains, where he became a legend. Born March 17, 1804, in Richmond, Virginia, he became the ultimate mountain renaissance man, founding fur trading businesses and acting as a leading guide, scout, trapper, and explorer in a huge territory. In his prime (roughly from 1820 to 1850), he served as an intermediary sometimes between native tribes and white government and settlers. In the winter of 1824–1825, he became arguably the first European to see the Great Salt Lake. (Now most believe it was likely Étienne Provost, but Bridger did carry that honor for a while.)

He was not only known for his accomplishments, skills, and experience, but also for his tall tales and his overall charm. During his lifetime he married two Native American women, one Flathead with whom he had three children and, after her death, the daughter of a Shoshone chief. He was conversant in many native languages, as well as English, Spanish, and French. He led troops in various Indian wars and discovered ways to improve and shorten the Oregon Trail. Suffering from numerous old-age ailments, Bridger peaced out July 17, 1881 near Kansas City, Missouri. Amazingly, this man traveled more in the 1800s than most people travel today, even with access to modern planes, trains, and automobiles.

SINKS ARE TUBS

Peter Sinks is just one of the many sinks in this potholed region, which is actually a fairly common geological phenomenon along fault lines. Near the top of Logan Canyon, there is a double-fault system where the limestone-rich crust has been exposed to water seepage. As limestone is soluble in water, vast cave systems form in this rock over time. These caves grow so much that they collapse, creating what are today called "sinks."

Peter Sinks, the chilliest spot in Utah, is one such basin that is almost a full square mile in size. The shape of this basin and the physics of air densities keep this tub dramatically cooler than anything nearby. First, cold air is much denser than hot air and therefore settles to the bottom of any container; this land depression is the perfect catcher's mitt for such a cold lump of air. Second, in addition to the basin holding the cold air, it also blocks any wind, which might possibly churn the cold out of its resting spot. Visitors will notice an extreme temperature difference within just a few steps; be careful testing this around the sinks, though, as drop-offs can sometimes be abrupt.

Just less than 30 miles from Logan, **Bear Lake** suddenly come into view on the left-hand side of the road. This massive, almost cyan-colored, Caribbean-looking body of water occupies a huge portion of the valley below. Bear Lake, the second-largest freshwater lake (behind Utah Lake, near Provo) is an elongated body of water, 18 miles long and 7 miles at its longest and widest points, it almost equally straddles Idaho and Utah. The original name, Black Bear Lake, given by its first European "discoverer," Donald Mackenzie of the Northwest Fur Company, was later shortened.

In the 1820s and '30s, this huge body of freshwater was a common meeting place for mountain men such as Jim Bridger and Jedadiah Smith, who traded goods amongst themselves and with Native Americans. Today this heritage is still remembered with the annual **Mountain Man Rendezvous.** Taking place in the middle of every September at **Rendezvous Beach,** this extended weekend festival is quite a spectacle with reenactments in period dress, as well as food and crafts made in essence of those days.

Garden City's outskirts immediately begin to dot US 89 as it makes its long descent into **Bear Lake Valley.** This town is rather famous for its raspberries. And it's got it all: raspberry baked goods, confections, and even a festival—the **Bear Lake Raspberry Days.** Though many, many restaurants serve raspberry shakes all year long, those visiting the first weekend of

August (Thursday through Saturday) should take advantage of the chance to experience this celebration of the annual, usually three-week-long harvest. In addition to special food treats, there is a parade, rodeo, beauty pageant, craft booths, music, and fireworks.

Garden City marks the end of this drive. The easiest way to return to the Salt Lake City area is to retrace the route, back along US 89, through Logan, and on US 89/91 to I-15. Those with a bit more time (or aiming for a northern destination) should consider making a loop up into Idaho.

Continuing north on US 89 leads to ID 36. This scenic byway ascends back west into mountain canyons that lead toward **Preston, Idaho**—the setting of the famous 2004 film, *Napoleon Dynamite*. Similar in scenery to the Logan Canyon drive, this route heads through much less developed land, on a smaller highway. Preston sits right on US 91; this north-south-running highway either leads straight back south toward Logan, Utah, or meets up with I-15, north of Preston and fairly near to **Pocatello, Idaho.**

BEAR LAKE, TIMES TWO

Salt Lake Valley isn't the only one that was underwater during the most recent ice ages. During the time that **Lake Bonneville** filled the Salt Lake Valley (up until about 12,000 years ago), Bear Lake's waters, too, were way above the current shoreline, and was about twice its current size. Though the two lakes were not connected, the Bear River actually flowed into Lake Bonneville at its largest point, and is thought to be responsible for the lake's final demise.

Lake Bonneville, a massive, prehistoric body of water had actually filled and drained many times during the ice ages. On its last time around, Boneville was growing and growing, fed by a moist climate and trapped in the **Great Basin.** At a certain point, it became large enough to accept the **Bear River.** This river drains a huge basin centered around the far northeastern corner of Utah's panhandle, and must have significantly increased the volume in Lake Bonneville.

When Bonneville acquired the Bear River's inflow, the lake began filling much more rapidly than before. The land at **Red Rock Pass** (near I-15 and the Utah–Idaho border) suddenly gave way, sending a catastrophic flood of truly epic proportions across **Idaho,** and through the **Snake River Valley.** This wall of water was more than 400 feet tall, traveled more than 70 miles per hour, and had a flow of 15 million cubic feet per second. This flood is responsible for many features of southern Idaho's Snake River Plane, including the out-of-place, rounded boulders now scattered across it.

IN THE AREA

Accommodations

Beaver Creek Lodge, 12800 US 89, Garden City. Moderate to expensive, rates vary significantly per season, length of stay, and activities. Call 435-946-4485 or 1-800-946-4485. Web site: www.beavercreeklodge.com.

University Inn, 4300 Old Main Hill, Logan. Moderately priced. Historic, downtown, hotel-style lodging. Call 435-797-0017 or 1-800-231-5634. Web site: www.uicc.usu.edu.

Attractions and Recreation

Bear Lake State Park. Administering the public beaches, picnic areas, campsites, and general recreation at Bear Lake. day-use and camping fees collected. Call 435-946-3343 or 1-877-887-2757. Web site: www.state parks.utah.gov.

Beaver Mountain, 1351 East 700 North, Logan. Winter and summer resort. Call 435-753-4822. Web site: www.skithebeav.com.

Dining

Indian Oven, 130 North Main, Logan. Moderately priced. Casual, but pleasant atmosphere; authentic Indian cuisine. Extensive menu with vegetarian options. Open for lunch and dinner, Monday through Saturday; closed Sunday. Call 435-787-1757. Web site: www.indianovenutah.com.

Le Nonne, 129 North 100 East, Logan. Moderate to expensive. Formal, upscale restaurant in renovated historic home; traditional Italian cuisine. Lunch Thursday and Friday only; dinner Monday through Saturday; closed between meals (call for hours). Reservations recommended. Call 435-752-9577. Web site: www.lenonne.com.

Other Contacts

Bear Lake Raspberry Days, downtown Garden City. Taking place the first weekend of August (Thursday through Saturday). A three-day festival with parade, rodeo, crafts, music, fireworks, and more. Call the Bear Lake Convention and Visitors Bureau: 1-800-448-2327. Web site: www .bearlake.org.

Cache Valley Visitors Bureau, (inside the Cache County Courthouse), 199 Main Street, Logan. Call 435-755-1890 or 1-800-882-4433. Web site: www.tourcachevalley.com.

Mountain Man Rendezvous, at Bear Lake's Rendezvous Beach; 2 miles north of Laketown, off UT 30. A one-weekend event in the middle of every September. Call Bear Lake State Park: 435-946-3343 or 1-877-887-2757. Web site: www.stateparks.utah.gov.

Stokes Nature Center, 2696 East US 89, Logan. Open Wednesday through Saturday 10 AM–4 PM, or by appointment. Check ahead for holiday and seasonal closures. Call 435-755-3239. Web site: www.logan nature.org.

Logan Ranger District Visitor Center, 1500 East US 89, Logan. Open Monday through Friday 8 AM–4:30 PM, all year. Call 435-755-3620.

Top of Utah Marathon. Taking place each year in the middle of September, on a Saturday. Call 435-797-2638. Web site: www.topofutahmarathon .com.

Utah State University, 1400 Old Main Hill, Logan. Call 435-797-1000. Web site: www.usu.edu.

Horseback Riding in the Church Meadows of the Uinta Mountains

CHAPTER

5

Mirror Lake Scenic Byway

Estimated length: Roughly 77 miles from Kamas, Utah, to Evanston, Wyoming

Estimated time: Half day (with return trip)

Getting There: Kamas, Utah, is a pretty straightforward town to reach. If coming from Salt Lake City, take I-80 east to its junction with US 40, just east of Park City. Head south a little more than 3 miles on US 40, taking exit 4 onto eastbound UT 248 toward Kamas. Travel a little more than 11 miles on this state highway, passing Jordanelle Reservoir to the south. Once in Kamas, a 0.2-mile dogleg to the north leads immediately to the well signed Mirror Lake Highway (UT 150) that departs eastwardly from the center of town.

    Those coming from the north or east also arrive via I-80. Begin by traveling south-southwest along I-80, taking exit 155 for Wanship/UT 32. Follow signs through this small town for UT 32, and continue onward about 16 miles southeast toward Kamas. The Mirror Lake Highway departs in an easterly direction from town, and is an extension of East Center Street.

Highlights: The **Uinta Mountains** are the tallest range in Utah. More than 100 miles long, they have more than 10 peaks taller than 13,000 feet and, aside from Alaska's Brooks Range, are the most outstanding east-west-running range in the nation. And, in a state with some of the nation's most notable scenery, UT 150 is one of the most unique drives. On a drive just 77 miles from Kamas to Evanston, visitors climb more than 4,200 vertical feet from Kamas to an elevation of 10,715 feet at Bald Mountain Pass without ever leaving the main road. Gaining height and losing heat, the ground cover transforms dramatically from grassy valley floor, to thick forest, to above-tree-line talus.

As this is clearly a high-elevation, mountain road, it is most certainly closed in winter—and can remain buried under snow until well into June. During summer, these mountains can endure some of the wintriest storms—at the same time that Salt Lake City and Moab are experiencing three-digit highs. Even in August, when the snow has left the range, thunderstorms occur almost like clockwork every afternoon; hiking and other activities are best done in the morning.

The Mirror Lake Highway (UT 150) begins in Kamas and cuts a crescent-shaped route northeast through the western corner of the Uinta Mountains, arriving at Evanston, just north of the Utah–Wyoming border. On its 56-mile journey, UT 150 ascends into Utah's tallest range, the Uinta Mountains, and then turns into WY 150 for another 21 miles before arriving in Evanston.

Uintah County, where the range sits, got its name from the resident tribe, the Ute and Uinta. Though Kamas is less than an hour from Salt Lake City—about as far away as Provo—this region remains largely unpopulated and is characterized by Indian reservations, one-stoplight towns, and ranches. Even if they wanted to, people couldn't feasibly live in the Uinta mountains year-round, as they sit under thick snowpack sometimes as much as nine or ten months a year. The highest peaks in this range stand more than 1.5 miles higher than the valleys below them.

Geologically speaking, the Uinta Range belongs to the Colorado Plateau. This tectonically textured and extremely colorful landmass occupies a Montana-sized region centered on the junction of Colorado, Utah, Arizona, and New Mexico, and whose watersheds feed the Colorado River.

Quite senior to mountain ranges in the south of Utah—such as the volcanic, 24-million-year old Henry and La Sal mountains—the Uintas are estimated to be roughly 60 million years old, created by tectonic uplift. The fundamental rocks found in this range—slate, shale, and, most visibly, quartzite—are dated to be 700–800 million years of age.

The region's human history begins in the valleys at the skirts of this giant range. With an elevation roughly in the 5,000- to 6,500-foot range, these relative lowlands are predictably more temperate than the peaks above. Atmospherically, they are actually quite dry, receiving less than 10 inches of precipitation annually. However these valleys do receive massive amounts of runoff from the Uinta Mountains, as well as from the Wasatch Range to the west—runoff fed by snowmelt and plentiful summer thunderstorms.

Because of the relative livability of the region's valleys, people have occupied the area for as long as 12 millennia. The Fremont were a northern offshoot of the Anasazi; a sub-group of the Fremont, called the Uinta Fremont (ancestors to the more modern Uinta Tribe), occupied this region initially. This hunter-gatherer people flourished with access to the area's abundant resources, particularly streams and lakes. With plentiful water flow comes many other things: elevated plant and animal life, as well as bountiful fishing. In addition the plentiful springs and streams is a natural abundance of glacial lakes—more than 1,000. During the cool, high-elevation summer conditions, these have always been great fisheries—then and today.

Fishing is far from the only recreational activity in these peaks. Also popular in modern times is camping, backpacking, road biking, and rock climbing. Though much too cold and snowy for the bulk of the year, these mountains provide a fabulous summer retreat from the Utah summer inferno—one of the only places within a day's drive of Salt Lake City where a sweatshirt can actually be worn in August!

Into the entire range protrudes but one major road—and this is the **Mirror Lake Scenic Byway** or UT 150, which makes a crescent-shaped sojourn through the Uintas' upper northwestern corner. Undeniably one of Utah's most scenic, it offers quite a change of pace from the plentiful red rock routes in the southern end of the state. The gateway to Utah's best summer escape, this highway can become inundated with vehicles during July and August. If the option exists, weekdays are a better choice for peaceful recreation. But if weekdays must be spent at work, crowds can be avoided by simply heading deeper into the backcountry.

If hungry in Kamas, consider a stop at **Dick's Drive In** before heading into the mountains. This retro burger joint is one of the most beloved in town, serving juicy grilled patties and thick shakes. If burgers don't sound appetizing, or vegetarian options are a must, visit **Pasillas.** This fresh-Mex café serves filling but less heavy dishes. Indoor and outdoor seating available during appropriate season. Located right on Main Street, this place requires no detour to visit. And finally, for home-style pizza of generous proportions, check out the **Summit Inn.** For dessert, they serve ice cream as well.

Leaving Kamas, the highway begins climbing immediately; after all, it has no choice if it is to gain so much elevation in its finite length. Before driving out of the valley and into the mountains, consider a stop at the **Kamas Ranger District Office.** In such a mountainous region, up-to-date information on road and trail conditions, as well as other updates pertaining to Mother Nature's alterations, can be quite helpful.

Just less than 5 miles east of Kamas is the small town of **Samak,** elevation 6,949 feet. A small town, just 4.7 square miles and 161 residents, its most noticeable feature is a big, bright sign advertising BEAVER CREEK NUDIST RANCH. This alleged colony is named after the stream running alongside the road until beyond the **Beaver Creek, Taylor's Fork,** and **Shingle Creek campgrounds.** (After these, the road passes the Beaver Creek headwaters and moves into the **Provo River Basin.**) Though this sign doesn't actually mark the location of a nudist ranch, it is a great, long-standing

---

EXPLORING THE UINTAS AWAY FROM UT 150

As mentioned before, this mountain range extends well beyond the Mirror Lake Scenic Byway—which merely teases at the northwestern portion of this massive mountain chain. Though only one road penetrates the high mountains, many skirt its slopes. Opposite UT 150, in the northeastern corner of the Uintas, stands the **Flaming Gorge Reservoir.** This dammed section of the **Green River** is popular for boating and other types of water recreation—but the real gem exists just downstream from the dam. Here, beneath the spillways, can be found some of the best fly fishing in Utah.

Downstream of the reservoir, is the **Uintah and Ouray Reservation.** This largely occupies the southern slopes of the Uintas, and contains a speckling of small townships along US 191 and 40. The largest of these, **Vernal,** has a population just more than 6,600, and from there, the town sizes fall off rapidly—with **Fort Duchesne** coming in at around 200.

The northern slopes of the range, in Wyoming, hold a lonely collection of tiny towns—by and large farming and ranching communities. Well off any kind of beaten path, these offer almost nothing to tourists.

A Classic Meadow Hike in the High Uintas

prank on passers-by. In a generally conservative state, this definitely causes people to pull over, and look more than once for the ranch. Many people even get out of the car and pose next to the sign along the way.

More than the home of fake nudist colony facades, Samak is also the location of the **Samak Smoke House & Country Store.** A quintessential smoke house indeed, this produces some fantastic, all-natural jerky made of turkey and beef. Its country store sells gift baskets with edible treats and goods made in the kitchen: biscotti, cookies, various coffees, nuts, and the like. These provide good finger food for camping trips, as well as gifts to bring back after a trip.

Heading northeast and upward, it is not even a mile before UT 150 leads into the **Uinta National Forest.** As national forests usually are, this one is filled with campgrounds, picnic areas, and trailheads. Along the way, drivers pass dozens of major pullouts and numerous other access points in the Ashley and Uinta-Wasatch-Cache national forests. In order to maintain these recreational sites, the Forest Service collects a fee from anyone who parks anywhere along UT 150 (or its side roads), payable at the ranger booth east of Kamas. The passes are also good for the **Alpine Scenic Loop** (chapter 3), east of Provo.

Immediately, campground access roads begin passing by. For those planning to tent here in shoulder seasons or chilly weather, these present the warmest options, as they are literally thousands of feet lower in elevation than many of the upcoming campgrounds. The first of these, **Yellow Pine Campground,** is also the location of a trailhead. Named for the creek it follows, the **Yellow Pine Trail** leads up quite high into the mountains, eventually reaching **Castle Peak.** Because

HIGH UINTAS CLASSIC: "NO WIMPS AND NO WHINERS!"

Every year during the middle of June, this two-day stage race sees beckons a few hundred road bikers. Since 1989, this summer solstice event has taken place almost immediately after UT 150 has been cleared of snow, with a sizeable cash purse available for licensed racers. One of the longest continuously held races in the United States, this begins—just as this drive—in Kamas and ends in Evanston, going right up and over Bald Mountain Pass, at more than 10,700 feet elevation, requiring an aggregate climb of nearly a vertical mile. Whether a cyclist or not, it's interesting to consider touring this route on a bike. Use caution when driving this road as this beautiful ribbon of pavement attracts many riders throughout the summer, whether or not it's race weekend.

of its elevation gain (more than 3,000 feet), it passes through a number of different ecosystems—but because of its long length (just more than 5 miles each way), the trail enjoys moderate grade.

About 8.5 miles east of Kamas, Upper Setting Road (Forest Service Road 034) intersects with UT 150. This high-quality, graded dirt road heads north and up along **Co-op Creek** toward **Castle Lake**. This road is actually one of the only means in the area to access this western Uinta high country without hiking the entire distance. From the terminus of the road, a trail heads west and joins up with the Yellow Pine Trail, mentioned above.

Soapstone Basin Road comes up just less than 6 miles later, this time heading south into the range. Located here is yet another campground, as well as another trailhead whose path branches into a spider web of options that trace along the bottom of **Soapstone Creek** and along the base of **Soapstone Mountain**. Hiking to the south, then East, and over a pass, will take hikers into **Mill Hollow**, and onto Mill Hollow Road (map recommended). This road parallels the south fork of the Provo River and heads northwest (parallel to UT 150) toward the towns of **Francis** and **Woodland**. For long distance hikers or mountain bikers, this very gentle 10.5 mile loop (with optional 3-mile-long dogleg) requires only about 1,500 feet of climbing on soft-packed soil—providing a very pleasant opportunity for extended exercise.

2010 HELICOPTER CRASH, NO CASUALTIES

Soapstone Basin was the location of a Coast Guard helicopter crash on March 3, 2010. Of five persons on board the MH-60T Jayhawk, none was killed. The chopper was returning home after a security detail at the 2010 Olympic Winter Games in Vancouver, British Columbia. Blizzardy conditions at the time of the crash were reported, and are suspected to be a cause for the accident.

Continuing along UT 150, the elevation rises to 8,280 feet as the highway passes a few more campgrounds, **Shady Dell** and **Cobblerest**. Past these, UT 150 climbs quite a lot and enters a region visibly carved by bygone glaciers. As the road bends around an abrupt 90-degree turn (from north to east), it finally reaches **Bald Mountain Pass** at elevation 10,715 feet—just 29 miles northeast of Kamas. Despite the impressive elevation of these mountains, the roundness and smoothness of their basins and saddles is somewhat surprising. Not at all knife-blade ridges, the environs rather feel like a safe, sculpted island in the sky.

The following section of road contains the densest collection of camp-grounds, side roads, lakes, and trails. Look for several lakes—**Trial, Lilly, Lost, Buckeye, Blue,** and **Wall**—as well as many campgrounds—**Lost Creek, Trial Lake,** and **Lilly Lake.** On the eastern side of the road will be some of the most popular attractions of their kind, the **Upper Provo River Falls** and **Campground,** roughly 24 miles after leaving Kamas and just before **Washington Lake.** Best steer clear of these on any normal summer weekend.

GOLD DIGGIN'

It's a rare spot in the Rocky Mountains that wasn't affected by the 19th-century gold rush. Once the '49ers had scoured and all but extinguished the veins and streams of California, remaining prospectors turned inland to the mountains of Idaho, Colorado, Utah, and Montana—anywhere precious minerals and metals could possibly be found.

Legend has it that as far back as the 1500s gold was brought to this range on behalf of Montezuma for safe hiding. Mixed fables hold that secret caves, rich with this precious metal, were scattered throughout the Uintas. Lore even tells that outlaws from Butch Cassidy's Wild Bunch era stashed stolen gold bricks in **Nine Mile Canyon.** It is also believed—and backed by archaeological evidence—that Spanish miners operated outfits powered by enslaved Utes in the 1700s.

These tantalizing rumors have long sent prospectors and treasure hunters into the Uinta Mountains seeking the alleged veins and secret stashes—but with such a massive, rugged piece of land to cover, these searches were understandably almost always in vain. Dark rumors have it that not only do dense deposits and rich stashes of hidden gold exist in these hills, but that a curse strikes down anyone attempting to access or carry away the riches. Though evidence is generally lacking, many have reportedly vanished or suffered terrible deaths in their quest for Uinta gold.

One of the most famous and corroborated stories is that of Kentucky-born, California mining magnate Thomas Rhoads. During a life that progressed generally westward, this man joined the LDS Church in Illinois in 1835. Before the official Brigham Young–led migration to the Salt Lake Valley, Rhoads went on a church-sponsored reconnaissance mission west of the Rocky Mountains. His party made it all the way to California, actually just ahead of the famously ill-fated Donner Party. In California, they had great luck as some of the first to take part in California's great gold rush. But in 1849, Brigham Young summoned this party to come to Utah, the Mormons' new-founded Zion.

Rhoades indeed came to Utah and, with his four wives and three dozen children, enjoyed wealth several degrees greater than most pioneers, building a massive home immediately south of Temple Square in Salt Lake City. About this time, Brigham Young, then president of the LDS Church, appointed Rhoades to the duty of recovering gold from hidden mines in the Uintas.

Under control of the Ute tribe, these mines had apparently been in operation since the 18th century, when pre-Escalante Dominguez Spanish had been running them with Ute slave labor. With the Spanish now gone, the Utes, who had little use for the metal, agreed to allow just one person access to these mines. That person was to only remove as much from the mine as he could carry each day. If anyone else tried to enter, or that person attempted to remove more gold than this, he should be executed. Chief Walker (or Wakara) approved this agreement, but under the condition that the gold be used exclusively for LDS Church purposes.

Allegedly, Thomas Rhoades made regular trips to the mine. Each lasted about two weeks, and would render just more than sixty pounds of the metal. When serious illness rendered Thomas incapable of this duty, he was replaced by his son, Caleb. This agreement continued until after Chief Walker and his successor Arapeen died; Chief Tabby allegedly discontinued the agreement, forbidding Mormons to access the mine.

Caleb reportedly made a few stealth missions to the mine, but wanted support from the U.S. government. He attempted to get rights to the land, claiming he could settle the national debt in exchange for these privileges. But the U.S. government undoubtedly viewed this as a crackpot deal and sent their own companies into search the land. Though they indeed found evidence of the old Spanish mining operations, they reportedly never found the mother lode and, to this day, its location remains a mystery.

Today still, the Rhoades family asserts that much of the gold on the Salt Lake Temple's Angel Moroni, as well as interior plating of the temple, was from Thomas Rhoades's Uinta trips. And many still believe in a lingering curse on the mine; people who have ventured into the hills in search of it have reportedly vanished or perished. In any case, many historians believe that Mormon wealth is directly linked to this vanished Rhoades mine.

Continuing along, the road begins to descend slowly, though it lingers for a while on this island-in-the-sky. Passing Mirror Lake, trails depart from either side of the road, popular for horseback riding and backpacking. Some campgrounds, like **Butterfly Lake,** and the **Highline Trailhead** exist in the area, as well as numerous rock climbing crags, including the more

Stream Meandering through a Meadow Beneath Ruth Lake

popular **Ruth Lake** and **Stone Garden** cliffs, both of which can be found by hiking from the left of the road.

The road continues its descent, passing tall peaks on either side—including **Hayden Peak** to the east at 12,475 feet. Along the way, six more campgrounds line the pathway in the **Hayden Fork** drainage down and toward the **Bear River.** Numerous small lakes exist mostly on the western slope, accessible by trails only. Many paths depart from the highway, or from dirt spur roads, leading to the backcountry. If roadside attractions earlier along the route were too mobbed, consider turning off onto a random road at this point in the drive and begin exploring by foot power.

As the mountains here were sculpted by glaciers, many of these hiking trails trace along the bottom of high meadows, making for beautiful walks without strenuous elevation gain. Though often beneath the ridgeline, lightning hazards do exist. Storms occur virtually every summer day in these mountains. Despite clear morning skies, clouds move in like clockwork in the afternoon and can present a dangerous threat to hikers exposed on higher terrain.

About 4 miles south of the Wyoming border, this road leaves the National Forest, marking the end of the campgrounds. Evanston, Wyoming, marks the "end" of the route. At this point, two options exist: backtracking WY/UT 150, or taking a more expedited trip. The quickest way back south is via I-80 West, heading back toward **Park City** and **Salt**

ROCK CLIMBING: HIGH-ALTITUDE CRAGGING

Reputed among rock climbers as having one of the most active climbing communities in the nation, Salt Lake City is indeed home to a rock-hungry bunch. Though the dry, warm climate allows for great spring and autumn climbing (and even passable winter conditions), the summers are simply too hot for the sport in the lower-altitude canyons.

The Uintas have presented a perfect escape from the summer heat since route development skyrocketed here in the late 1990s. Though an occasional technical climb has been established here and there since the 1960s, routes were generally few and far between—and often not very difficult. However 1995 saw the first major developments for rock climbing in the Uintas, with traditional and sport climbing lines going up rapidly in its quartzite. The number of routes quickly went from less than 20, to more than 400—in little more than a decade. Today these climbs are spread among 20 main areas, and are detailed in Nathan Smith and Paul Tusting's guidebook, *Uinta Rock.*

Afternoon Thunderstorms Viewed from Amethyst Lake Trail

Lake City. Along the way, I-80 runs though **Echo Canyon**—the weakness in the Rocky Mountains through which the First Transcontinental Railroad ran. From Evanston, the highway miles back to Salt Lake City amount to about 80, requiring a bit more than an hour's drive.

IN THE AREA

Accommodations

Summit Inn, 80 South Main Street, Kamas. Affordable, casual, family-friendly pizza restaurant. Call 435-783-4453.

Dining

Dick's Drive In, 35 Center St, Kamas. Very affordable, very casual American cuisine. Call 435-783-4312.

Pasillas Restaurant, 185 South Main Street, Kamas. Affordable. Casual, fresh Mexican and southwestern. Call 435-783-6982.

Samak Smoke House & Country Store, 1937 East UT 150, Kamas. Name says it all. Call 435-783-4880. Web site: www.samaksmokehouse.com.

Other Contacts

Ashley National Forest Regional Office, 355 North Vernal Avenue, Vernal. This national forest presides over nearly 1.4 million acres in Utah and Wyoming, 180,000 of which are wilderness acres in the Uinta region. Call 435-789-1181. Web site: www.fs.fed.us/r4/ashley.

Bureau of Land Management Salt Lake: Northern Uintas Office, 440 West 200 South, Suite 500, Salt Lake City. Call 801-539-4001. Web site: www.blm.gov/ut.

Bureau of Land Management Vernal: Southern Uintas and Uinta Basin Office, 170 South 500 East, Vernal. Call 435-781-4400. Web site: www.blm.gov/ut.

Dinosaurland Travel Board, 55 East Main Street, Vernal. Regional visitor information; online listings for dining, recreation, lodging. Call 435-789-6932 or 1-800-477-5558. Web site: www.dinoland.com.

High Uintas Classic, bicycle race. Web site: www.evanstoncycling.org.

Kamas Ranger District Office, 50 East Center Street Kamas. Open Monday through Friday. Call 435-783-4338. Web site: www.fs.fed.us/r4/uwc.

National Scenic Byways Program. Call 1-800-429-9297. Web site: www.byways.org.

Uinta-Wasatch-Cache National Forest Office, 125 South State Street, Salt Lake City. A recently united Forest Service office, now overseeing 2.1 million acres of land; Web site updated with alerts. Call 801-236-4300. Web site: www.fs.fed.us.

Delicate Arch as Seen from View Trail in Arches National Park

CHAPTER

6

Moab to Arches

Estimated length: About 25 miles from Moab to the northernmost part of Arches (Devil's Garden); 65 miles roundtrip to drive each of the park's major, paved roads

Estimated time: Half a day for a car tour; a full day if taking a hike or two

Getting There: Start by finding Moab. (More information on getting to Moab in the "Getting There" section of chapter 8, Moab to Canyonlands' Needles.) Once there, the route to Arches couldn't be simpler. Head out of town in a northerly direction on US 191/Main Street, about 4.5 miles from the center of Moab. A well signed right-hand turn for Arches National Park departs from the highway to the east. Shortly after this turn comes the fee station, visitor center, and park entrance.

There is a weekly fee to enter the park; national parks passes are accepted. A "Local Passport" can be purchased here for regional parks and monuments: Canyonlands, Natural Bridges, and Hovenweep.

Highlights: Though difficult to imagine when driving US 191, **Arches National Park** sits immediately off this highway—not even five minutes north of Moab. Immediately after the fee station, the park road winds steeply up, away from the valley floor, and into one of the country's most unique natural rock gardens. Though named for its most unique feature, the arches, this park actually has several different regions, each with its own flavor. Never leaving the car, it is possible to see spires, fins, arches, and even petrified sand dunes. When in the **visitor center** near the park's entrance, pick up a free map; this will give an overview of the park's simple road system and the relative locations of the major attractions.

Arches National Park has more than 2,000 natural rock windows, giving it the densest concentration of such features in the world. In addition to that obvious attraction, there are all kinds of unusual rock shapes: from hoodoos to spires, and sheer cliffs to globular fins. Within the park live 52 known species of mammal, 186 of birds, and over 480 types of plant.

People first passed through this region roughly 10,000 year ago. These migratory, hunter-gatherer societies appeared first at the end of the last ice age. Within 2,000 years of today, the ancestors of the Freemont and Puebloan Indians made a part-time home here. The Ute Indians were present in the area when the first Europeans began their exploration of it. Contact with the Utes was documented by these early European explorers—but the Utes also left their own stories in the forms of pictographs and petroglyphs.

For obvious climatic reasons, however, these peoples were only able to live here during selective times of year, when water was naturally supplied by rain and snow. During summer, the region was largely uninhabitable, as nearly all streams run dry. On your drive, try to imagine what the area would be like without roads, water lines, or infrastructure of any kind—and you'll probably feel a great deal of respect for those who were able to live on this pre-"civilized" land.

The first permanent resident of the area, John Wesley Wolfe, brought his son along and built a cabin in 1889. Alone, they lived in this cabin until 1910. The remains of this structure are visible at the **Wolfe Ranch** today (located at the trailhead for the **Delicate Arch Trail**). But the first person to take specific interest in promoting the area's peculiar beauty was Alexander Ringhoffer, a prospector and miner. He summoned sufficient interest, and it was designated Arches National Monument by the federal government in 1929.

The main out-of-car activity in Arches is hiking; in fact, to even see the park's most famous landmark, **Delicate Arch,** a short hike is required. Therefore, visitors should prepare themselves with adequate rations before leaving town. Numerous hiking trails in the park range from very easy to strenuous. As this is a national park, trails are typically very well maintained; however severe weather may sometimes damage trails more quickly than the park service can rebuild them. Regardless of whether visitors intend to leave their cars or stay in them, they should all supply themselves with plenty of water, food, clothing, and sunblock—best done while still in **Moab.** (Outside of Moab, the last structure in the park is the visitor center, where very limited supplies are available.)

The largest store in town supplying groceries and beverages is the **City Market.** In addition to the obvious wares, it also has a small lunch bar, stocks pre-made sandwiches, and carries firewood bundles, and ice. For those requiring outdoor-specific gear, **Gearheads** has the most general selection, ranging from rock climbing to camping. One critical item to carry in the desert is sunblock. The park's elevation is roughly between 4,100 and 5,700 feet, and less than a third of its days each year are cloudy. Lightweight, light-colored, loose-fitting clothing helps a lot, too. Before departing on any kind of hike, be absolutely sure to have plenty of water or electrolyte beverage. Some suggest one liter per person, per hour of activity, but it doesn't hurt to have (a lot) more.

HIGHS IN THE 90S, LOWS BELOW FREEZING

Arches has a quintessential desert climate. This translates to dominant sunshine, bone dry air, and huge daily temperature swings. By day, the ground bakes in radiant heat, supplied by uninterrupted solar rays. But as soon as the sun goes down, the arid, thin air does very little to retain any of this heat, allowing the temperatures to plunge. Summer highs typically reach easily into the mid-90s, with overnight lows near 60 degrees F. Winter afternoons reach the low- to mid-60s, with nights getting down below 20 degrees F. Late November through mid February are typically the cloudiest months, rendering spring and autumn the most pleasant times here for heat-sensitive humans.

Heading north from Moab is simple and uneventful. Restaurant and hotel signs lining the roadside thin as US 191 leads out of town. The highway crosses over the historic Colorado River. Taking a right turn here would lead east on UT 128, and onto chapter 7's tour of the Moab–Colorado River–Castleton Valley–La Sal Mountain Loop Road tour.

Four and a half miles from the center of Moab comes the obvious right-hand turn off for Arches, and just more than 0.5 miles beyond that comes the park entrance and fee station where all must pay or present a national parks pass. As long as the vehicle is stopped here, it's worth taking a left and checking out the **visitor center.** In keeping with tradition, this is where the Park Service distributes all kinds of free and not-free literature: maps, books, guides. This is also the last outpost of civilization inside the park. Though Arches is stacked with rangers and other visitors, none of these people will be selling water or food.

As the road ascents steeply away from the visitor center, keep an eye open for a small pullout and placard on the right-hand side of the road (just before the road curves to the left). Those interested in the area's geology will get a clear, visual explanation of one of the primary causes for the park's strange formations. The **Moab Fault,** along which US 191 sits (between Moab and Arches), formed roughly six million years ago. When tectonic movements became too much for the earth's crust to handle, it cracked. Subsequently the eastern (Arches) side of the fault ended up a half a mile lower than the western side. This major vertical shift and subsequent rock fracturing is one of the contributing factors to the strange erosion patterns that have occurred in the park (as described in the "Geology" sidebar, above).

About 2 miles beyond this sign (and 2.5 miles uphill from the visitor

---

PECULIAR GEOLOGY

Roughly 300 million years ago, seas covered the area where Arches is today. Over time the waters retreated, then returned, then retreated and returned—multiple times—thus leaving behind significant salt deposits. Eventually the sea water left for good, and the residual, thick salt layer was gradually covered in other sediments: mud and sand. This would eventually become sandstone layers—primarily Navajo and Entrada—resting atop the solidified salt beneath.

Naturally the salt, under that much pressure, destabilized and could no longer support the layers above. The sandstone began slipping and sliding over the millennia, rising and falling on top of the uneven salt layer. As it did this, it splintered all over the place. Some pieces dipped low and some remained high. All of this tilting and fracturing, dipping and doming, left the rock vulnerable to some strange weathering patterns. Natural unevenness in the rock's density and solubility allowed for arches and hoodoos to form—as the softer rock beneath sturdier rock wore away more quickly. Water, ice, and wind each played their parts in taking the weaker rock material, yielding the spires, fins, and rounded shapes remaining today.

CRYPTOBIOTIC WHAT?

People foreign to the desert may not be aware that it is critical for hikers to stay on established trails—inside or outside of the park. Unlike in more lush environments, the desert's fragile ecosystem can literally be damaged by one person's footprints! As hot and dry climate cannot possibly support widespread vegetation, there are no root structures to hold everything in place. Rather, the soil itself is alive.

Cryptobiotic soil, as it's called, is a fragile mixture of lichens, mosses, and cyanobacteria—often invisible. This delicate carpet is extremely slow-growing, and in some places represents 70 percent of the desert's living surface. These organisms play many rolls in keeping the ecosystems healthy, from maintaining a chemical balance in the soil, to preventing devastating water and wind erosion. Without it, the desert could quite literally blow away.

center), the road comes to **Park Avenue,** a trailhead and viewpoint. From the parking lot, the namesake feature is visible. This collection of tall, thin, and dead-vertical fins stands clustered together to resemble a sandstone version of Manhattan. From here, it is possible do hike a 1-mile trail that descends down into the canyon beneath these fins, eventually leading to the **Courthouse Towers** parking area. The trail loses about 300 feet in elevation as it descends to the wash beneath the towers (almost always dry). Those making the hike to the cluster experience a particularly impressive view, being dwarfed by the tall, clustered fins in a way different than from the road.

On the way to the **Courthouse Towers** pullout, notice the **La Sal Mountains Viewpoint** on the right side of the road. The **La Sal Mountains** are a commanding range that soars far above Moab. (For a closer look, consider taking the Moab/Cas-

SISTER RANGES

Anyone who's seen the **Abajo** or **Henry Mountains** (also in southeastern Utah), has noticed an immediate resemblance to the La Sals. All impressively tall and distinct ranges, they rise high and alone above the surrounding Colorado Plateau desert. Each was formed roughly 25 million years ago by igneous intrusion. Magma forced itself high into the upper strata of earth's crust, cooling slowly beneath the surface. Covering layers of sandstone eroded away over time, exposing rock generally classified as porphyry—characterized by large crystals formed during the gradual, subterranean cooling of magma.

Petrified Sand Dunes David Sjöquist

tle Valley/La Sal Mountain Loop Ride, chapter 7.) These mountains rise far above anything around, topping out at 12,721 feet, at the peak of **Mount Peale**—almost 8,700 feet higher than Moab. An artistic contrast to the desert's oranges and reds, these deep blue peaks hold snow into the summer season, while the valleys below bake at triple-digit temperatures.

From the Courthouse Towers parking just down the road, sturdy, tall fins dominate the view. Thick and imposing, these are quite different than the arches coming up farther along. Rising hundreds of feet above the flat, broad valley below, they are an obvious high-rise simulacrum. Signs indicate **Sheep Rock, The Organ, Tower of Babel,** and **Three Gossips.** Looking closely, anyone should be able to spot Sheep Rock and the Three Gossips without any help. The Organ and Tower of Babel are not quite as intuitively recognized, but roadside placards help.

Leaving this tower metropolis, sit back and enjoy the general scenery, as the only official viewpoint in the next 4 miles is for the **Petrified Dunes.** Though not as visually striking as many of the park's features, the dunes are still worth considering. Bona fide sand dunes, these were covered over time by other layers of sediment and compressed into solid rock. Later these overlaying rock types eroded and leaving the dunes exposed. They can be recognized as the pale rock mounds poking out between clumps of low vegetation.

Motoring along, the next distinctive attraction is **Balanced Rock**— just 0.2 miles before the road splits in twain. It will be a sight to see when this unlikely formation finally topples! This vaguely egg-shaped boulder, roughly the size of three buses, rests atop a tapered, slanted rock pillar. The total height of this feature is 128 feet, 55 of which come from the show-piece itself. This suspicious-looking feature is not a strange man-made prank; rather it is a classic example of softer bedrock eroding beneath harder capstone. A short, paved path leads directly to the base. Worth noting: the trail to Balanced Rock and viewpoint trail for Delicate Arch are the only two handicapped-accessible trails in Arches National Park.

Just a few seconds' driving leads to the turnoff for the **Windows Section** of the park. This 2.5-mile road heads east through **Ham Rock, The Garden of Eden,** and **Cover Arch** to reach a parking area and viewpoint for the **North and South Windows, Turret Arch, Cove of Caves,** and **Double Arch.**

Windows Road

Though it's a bit of a stretch to call it **Ham Rock,** this first formation you see after taking the turnoff, is one of the most visible in the park—visible from many vantage points. Roughly shaped like a ham (…or egg…or avo-cado), it is perched high upon a Navajo sandstone dome, making it easy to spot. Just after Ham Rock is a turnout to the left. This overlooks the **Garden of Eden, Cove Arch,** and **Elephant Butte,** which stands 5,653 feet above sea level.

The end of this road is clearly marked with a turnaround and large parking area. From here, a generally flat, 0.4-mile trail heads from the northern end of the parking lot to **Double Arch,** providing a leg-stretch-er/vista combination. Though Double Arch is plainly visible from the road, the broad trail leads almost directly to the base of this remarkable feature, providing a much more impressive view. Those planning to get out for photographs or a hike should respect Arches' requirement that all cars park in the designated areas, lest traffic accumulate at this popular viewpoint and trailhead.

Departing eastwardly from the same parking area is a 0.5-mile-long trail to the **Windows and Turret arches.** Though visible from the road, these reward hikers with a much superior view than that had on pavement. The **North and South Windows arches** are actual two twin-ish windows in the same rock fin. **Turret Arch,** though relatively small, has a fairly unique look, with one primary arch reaching to the ground, and a secondary, smaller hole resembling a window. Taken in view with the host rock formation and the trail steps leading up to it, it quite resembles a house with a turret. To return to the main road, do an about-face and head back west 2.5 miles.

Hang a right and continue northeast along the main road, stopping at **Panorama Point.** From here, the landscape scoops down and away to reveal gently sweeping sagebrush-covered hills in the foreground with a backdrop of globular pinnacles and domes. From a distance, it is much easier to see the clear bands of pale yellow and rust-orange sandstone that had once been continuous strata—before wild tectonic uplift and subsequent erosive forces sculpted them into the distinctive art of Southeastern Utah.

Delicate Arch Road

Two and a half miles after returning to the main road, make a right hand turn if you want to see the most famous arch in the state, **Delicate Arch.** Though not the largest, this showpiece arch seen on the Utah license plate is arguably the most aesthetic. It stands alone atop an expansive, buffed sandstone dome devoid of vegetation and is a freestanding arch of more than 180 degrees. At one point, government officials even considered varnishing this arch to prevent further erosion and the arch's imminent demise—but plan was deemed contrary to the mission of national parks and eventually forgotten.

To see this arch, some walking is unavoidable. Some options exist: hike to the arch itself, or view it from afar. To see it from a distance, two separate viewpoints can be reached via quick-and-easy hikes. These yield somewhat distant, though strikingly scenic, vistas; on the other hand, a proper hike takes you directly to the arch itself. For those with the time to hike directly to the arch, the 1.5-mile journey would be worth the effort—even if it didn't lead to the Delicate Arch. Along the way, this path gains a bit elevation, but as passes through brush, across slickrock, and in washes, the

Balanced Rock David Sjöquist

Trail to Double Arch at the End of the Windows Road

effort is hardly noticeable. Finally emerging at the end of the hike, Delicate Arch appears, itself hidden from view until the very end of the hike. Its absence along the way is perhaps what allows visitors to truly enjoy the surrounding geology and wildlife for their own sake. (To pick up this trail, head to the **Wolfe Ranch.**)

If, however, there is not enough time to complete this trek, simply drive to the terminus of the Delicate Arch Road, from which two short viewpoint access trails depart. Both are casual (one is just 150 feet; the other is 0.5 miles). The lesser trail is one just two handicapped-accessible trails in the park (the other one being for Balanced Rock). Both viewpoints are separated from the arch and its rounded sandstone base by a deep and narrow canyon. This dramatic relief makes for quite a view, and renders Delicate Arch quite impressive in appearance.

Leave the Delicate Arch spur of the road and regain the main road again. To see the rest of the park, make a right-hand turn and head north. This portion of the park is much more pedestrian-accessed than the previous car-friendly sections. Though two roadside viewpoints come up quickly, most of the attractions ahead must be accessed on foot. That is not to say that the roadside scenery will disappoint; just many of the arches must be require some effort to peep—a bonus for some!

This next section of Arches is called the **Fiery Furnace.** Just more than a mile from the junction with Delicate Arch Road comes the **Salt Valley Overlook** on the right. Salt Valley is the place in which the John Wesley Wolfe ran his small cattle ranch. Here the pale yellow and dark red sandstones have crumbled into buttes of dirt and talus, upon which grow twisted desert scrub. Take in this view, because the next vista has a completely different aesthetic.

Almost immediately after this pullout is the **Fiery Furnace Viewpoint,** showcasing a comical-looking collection of bubbly, knobby spires. The predominant rock color here is a deep red, and there are rows and rows of these pillars—not a stretch to say that it resembles an inferno in the belly of this round, broad valley.

WOLFE RANCH

In 1898, John Wesley Wolfe and his son, Fred, relocated from Ohio to Utah. Prior to the move, John had received a serious leg injury during his service in the Civil War. Seeking a warm, dry climate to soothe his persistent aches, he and his son set up shop on 100 acres at this location in the Salt Valley Wash. Here there was just enough of a water supply and corresponding grasslands to establish a very modest ranch.

John and Fred remained at that spot until returning to Etna, Ohio, shortly before John's death in 1913. To this day, the original log cabin, wooden fences, root cellar, corral, and other wooden structures remain—largely thanks to the arid climate in which decay progresses at an extremely slow rate. Today, the doors of these structures remain open for viewing's sake.

Those up for a five-minute hike can take a walk to view even older human artifacts from the in same locale. A fairly dense collection of **Ute petroglyphs** remain etched on boulders located in a wash, well signed from the same parking area as for the Wolfe Ranch. Depicted are several riders on horseback, as well as bighorn sheep and a few smaller animals.

Sand Dune Arch can be reached by a couple of different paths; the shortest trail to it departs from a pullout on the right hand side of the road, about 4.5 miles past the Windows Road junction. This trail reaches the arch in just a quarter of a mile. Surrounded by other rocks, sheltered, and often shady, this has a much different feel than many of the other arches. Sand Dune Arch rises out of a bed of deep sand (hence the name) and has an almost jungle-gym feel. Kids particularly like this one, as they feel as if they're in a larger-than-life sandbox.

For more of a hike, continue along this same trail. It leads out of the rock fin maze, through sagebrush, and to Broken Arch in just another mile. (This trail system is also accessible by a shorter approach from Devil's Campground site #40; the route is easily seen by looking at the Arches

PETROGLYPHS OR PICTOGRAMS?

These two words are often thrown around interchangeably. And they're used all the time in southern Utah, as the dry climate and abundant rock provided pristine lasting tablets to the artists of times past. Visible in Arches National Park is the work of the **Ute** and **ancestral Pueblan** tribes. Some of southwest Utah's rock art dates back roughly 2,000 years, spanning many different eras of these peoples. Much of the work can be attributed to shamans. After self-inducing hallucinogenic states, they were able to communicate with the spirit world. These experiences would then be recorded on the rock.

There is a simple difference between the two terms, worth learning: a **pictogram** (also called **pictograph**) is an image painted on the rock with any number of available dyes and pigments. Back in the days before latex paint and Red 40 other coloring agents were used such as animal blood and plant pigments.

A **petroglyph** (examples of which can be accessed by a trail at the Wolfe Ranch) is an image pecked or carved into the stone's surface. On Utah's sandstone, this renders a striking visual effect as the deep red varnish on the rock's surface can be chipped away to expose much lighter, less oxidized rock below. Over the centuries, artists used antlers, bones, harder stones, and later metal tools to create these.

One rough relative age indicator of petroglyphs is the etching's coloring. The newer images of a set will be much lighter than the older ones, as the manually exposed rock has simply had less exposure to the elements and therefore less time to oxidize. Depending on the climate, petroglyphs have the potential to remain in visible much longer than pictographs.

Climb from Visitors Center into the Park

National Park Map, available in the visitor center.) Broken Arch isn't actually broken at all, but by appearances it looks as if two arms of rock have leaned over to meet each other.

Dirt Road Spur

The next turnoff (on the left hand side) is to a dirt road forking off to the left (just a mile before the terminus of the main road). This spur tours the Salt Valley before forking in two. Continue straight all the way to the park's border (and beyond), or to see more arches, hang a left and drive another 1.1 miles to a pit toilet at the **Klondike Bluffs** area. From this parking area, a pedestrian leads to **Marching Men** and **Tower Arch.**

This section of park is the sweet spot for four-wheel-drive high-clearance vehicles. From the main dirt road, an 11-mile strip of challenging dirt works back south toward Balanced Rock and a few other points of interest. Before attempting these roads, it is best you consult with a good map, a weather report, and a park ranger to be sure there are no park-enforced closures or nature-induced impasses.

Back on the main road, a quick drive to the north delivers the **Skyline Arch** pullout. A hiking trail does lead to the base of this arch, but unlike most, this one is actually better viewed from the road. From afar, it appears most impressive, as it crowns a tall rock fin. But, due to topography, it all but disappears from view to those standing directly beneath. A massive rock fall in 1940 roughly doubled its size to its current span of 69 feet.

The **Devil's Garden Trailhead** area is the very last stop on the road. From this little hub are many attractions. There is the **Devil's Garden Trail** itself, a dirt road leading to a campground, and yet another trailhead leading out from the campground. There is even potable water available at the campground—so don't be shy about refilling.

The **Devil's Garden Campground** is the only campground within the park, and quite a pleasant one. Surrounded by a cluster of house-sized rock domes and filled with juniper, it has a cozy feel. Much more sheltered than other sections of the park, it is the ideal location for Arch's only campground. The setup here is semi-primitive, with groomed sites, picnic tables, grills, and toilets. Though drinking water is available, there are no showers; a few sites are large enough to accommodate RVs. There is an extra fee per night for the sites. Reservations are possible online or by telephone,

---

EDWARD ABBEY COUNTRY

It's possible that you could have escaped high school without reading one of Edward Abbey's books—*The Monkey Wrench Gang, Desert Solitaire, Fire on the Mountain,* or others from his lengthy list of titles—but it would be unfortunate. Abbey was a die-hard environmentalist and made a lifelong career of writing fiction and nonfiction on his favorite subject, the **Four Corners Region** of the American Southwest.

During the late 1950s, Abbey worked as a ranger in Arches (then **Arches National Monument**) where he wrote the notes for what would become *Desert Solitaire* in 1968. Now a mountain biking, rock climbing, river rafting, and general outdoorsy hub, Moab was, at that time, a bleak uranium mining center. Visible traces of the old mining days are still scattered throughout the area today.

March through October, for an additional charge. To hike from the camp-ground, find site #40 and depart to the south, traveling along a loop-and-hook-shaped trail past **Broken Arch** and **Sand Dune Arch.** (This is the same trail as is accessed by the Sand Dune Arch pullout on the main road, above.)

IN THE AREA

Arches National Park, P.O. Box 907, Moab. Call 435-719-2299. Web site: www.nps.gov/arch.

City Market, 425 South Main Street, Moab. Call 435-259-5181. Open daily. Web site: www.citymarket.com.

Devil's Garden Campground Reservations. Extra fee required to reserve a campsite. Call 518-885-3639, 1-877-444-6777, or 1-877-833-6777 (TDD). Web site: www.recreation.gov.

Gear Heads, 471 South Main Street, Moab. Call 435-259-4327 . Open daily.

Ascending into the La Sal Mountains on the La Sal Mountain Loop Road

CHAPTER

7

Moab, Colorado River, Castle Valley, La Sal Mountain Loop

Estimated length: About 63 miles (or 75 miles with the Fisher Towers detour)

Estimated time: A short half day unless stopping to hike, wine taste, or rock climb

Getting There: To get to Moab, take I-70 to exit 182, turning south on US 191. Thirty-two miles later, this turns seamlessly into Moab's Main Street. (For more specific instructions on how to come from Salt Lake City, see the "Getting There" section in chapter 8.)

Highlights: This loop drive departs from the world-famous, outdoorsy town of **Moab,** tours along the **Colorado River,** past a winery, the famous sandstone towers of **Castle Valley,** and eventually up and into the sub-alpine, too-fun-to-drive **La Sal Mountain Loop.** This route starts and finishes right in Moab, and is a succinct way to tour some of the area's more diverse terrain.

As always when leaving town in the desert, be sure to do a quick inventory of water, food, and gas. Though this route never gets terribly far from Moab, it has virtually no services along the way. Once satisfied with the trip's rations, point the wheels north and leave **Moab** by heading north on

Main Street/US 191. Just 3.5 miles from the center of Moab (and hardly a mile from the outskirts), make a right onto UT 128, just before the bridge over the Colorado River. When making this turn, reset the trip odometer. In the twists and turns of this narrow canyon, hiking trails, pullouts, and other features pop up rapidly and are often hard to spot, even by those who know their location along the roadside.

Until 2009, there was a freshwater spring running from a pipe fitted into the sandstone walls on the south side of this road, just a quarter a mile past the turnoff. Called **Matrimony Spring,** it allegedly gave couples great marital luck. This water was untreated by any chemicals and flowed around the clock—so was an excellent water source for campers, 24/7. However, obligatory testing by the City of Moab deemed the waters unsatisfactory for drinking, sadly bringing this long-term tradition to an end. Perhaps one day the spring will reopen for use, and couples can hope for happy futures again.

Continuing roughly east, keep an eye peeled for **Updraft Arch.** A pullout just 0.3 miles past Matrimony Spring provides safe viewing of this window. This eye-shaped slit in the already impressive Navajo sandstone cliffs high above the road is located just to the east of the pullout.

Not even 2 miles after the start of UT 128 begins the **Colorado Riverway Recreation Area.** Operated by the Bureau of Land Management, the area features several riverside campgrounds and picnic areas, as well as trails for hiking, mountain biking, and off-roading, and open lands for wild camping. Its developed campgrounds stretch quite a while along the banks of the Colorado River. Though sites are primitive (read: no electricity or running water), the campgrounds' views, awesome locations, and tidiness render them quite popular. Along UT 128 are seven such campgrounds, five of which are clustered around **Big Bend,** about 6.5 miles east of US 191. Among these are roughly 100 sites. (An overview of the campgrounds is posted on a BLM kiosk just one mile onto UT 128.)

Over the years, many of the rivers and mountains in the United States have been renamed so as to fit more properly into modern vocabulary. Well—this one still hasn't been: **Negro Bill Canyon.** Named after William Granstaff, this shady, narrow canyon is one of the few around with a perennial stream. The running water is usually a bit high in the early spring and dips a bit lower during the late spring and early summer months. Even on warmer days it still refrigerates the already cool canyon.

A 2.5-mile-hike along the canyon bottom leads to the **Morning Glory**

ELK MOUNTAIN MISSION, WILLIAM GRANSTAFF AND THE SAGEBRUSH REBELLION

In May 1855, after a very rugged and topographically challenging journey from Salt Lake City (founded in 1848), a 41-man group of missionaries arrived in the Spanish Valley and began to set up shop. Their goal was to meet and convert local peoples to Mormonism.

They erected a stone fort called **Elk Mountain Mission** at the northern end of present day Moab—actually where the **Motel 6** parking lot is today. They planted crops and attempted to establish friendly contract with local Ute tribe, who inhabited southern Utah. These Mormons perceived their arrival as successful, as they had already baptized more than a dozen Utes by the middle of July.

The Utes, however, didn't like what they were smelling, and trouble between the two groups started to brew. The Utes grew mistrustful of the valley's new conversion-happy residents. According to the Mormons' September 20 journal entries, crops from the fort were being stolen at night by Utes. On September 23, one Mormon man was allegedly shot by a son of the Ute chief. This sparked a gun battle at the fort where a handful of Ute attackers were killed—meanwhile other vicinity ambushes knocked off a few Mormons. Finally, the Utes torched the Mormon crops, giving the fort residents the final ousting they needed. The following day, all surviving Mormons packed up what they could carry and headed north, shy a few dozen cattle and five horses.

This fort would remain abandoned for more than 20 years until a trapper and cattle driver, William Granstaff, and another trapper known only as "Frenchie," would come to occupy it. William came to the region in the 1870s as prospecting and cattle-running cowboy. The namesake of Negro Bill Canyon, Granstaff grazed his cattle there regularly. In 1877 he and a cohort helped themselves to residence of the Elk Mountain Mission Fort in Spanish Valley—where Moab is today. But their residence wouldn't last long; Frenchie disappeared from records and in 1881, Granstaff was accused with bootlegging booze and hightailed out of the area.

Arch. Though the terrain is relatively flat, running water can require some skilled stone-stepping, making the roundtrip journey up to two or three hours long. Because stream crossing is required, it's not unreasonable to consider doing the walk in some beefy sandals, lest the water wet and stink a pair of good shoes and create a blister patch. Along the way, a handful of side canyons branch off—watch for a small sign about 2 miles from the

View from Castle Creek Winery

trailhead that points to Morning Glory Natural Bridge. In about a half a mile, the path leads directly to this 243-foot-wide arch.

Back on UT 128, the next 9 miles of driving lead past several campgrounds and day-use areas: picnic grounds, boat put-ins and take-outs. Many are concentrated around the Big Bend portion of this river where the Colorado makes a sweeping S-curve. At this location is a smattering of obvious boulders on which rock climbers will be playing in good weather. It might be possible to spot climbers higher up, on the canyon rim walls, as a few multipitch rock climbs ascend these cliffs.

Just before turning off UT 128, into Castle Valley, look to the north for the obvious **Castle Creek Winery** and adjacent **Red Cliffs Lodge**. Together the two operations occupy a spacious and green campus on the northern side of the Colorado River. Castle Creek, Utah's first and largest commercial winery, was opened in 2002. Now offering four whites and four reds, it produces more than 15,000 gallons of the goodness each year. Planting its own cabernet sauvignon and syrah in 2006, the winery was able to harvest and press some of its own grapes in 2008. Castle Creek has won multiple awards for several of its wines, from the Finger Lakes and Pacific Rim international competitions, to the Utah Best of State Award in 2009. Today it offers indoor and outdoor dining, tours, and wine tasting to anyone passing through.

Red Cliffs Lodge offers guest beds, as well as onsite and backcountry recreational opportunities—somewhat like summer camp for the whole family. Choices for accommodations include 20 cabins and 80 rooms, all overlooking the Colorado River. Packages include river rafting, horseback riding, scenic flights, golfing, jet-boat rides, off-roading, mountain biking, and hiking. Just on the grounds, guest have access to a swimming pool, fitness facility, tennis courts, and small museum (see sidebar, below). Ideal for special occasions and team-building activity, this lodge can attract large parties. So those wishing to avoid the company of large wedding parties should inquire when making reservations.

Very shortly after Castle Creek and the Red Rock Ranch (and about 15.5 miles east of US 191), make a right turn and head south into Castle

---

MOAB MUSEUM OF FILM & WESTERN HERITAGE

Even those visiting the area for the first time will surely recognize many of its vistas simply because of the many famous photographs and films shot on location here. Castle Valley and the Colorado River's iconic scenery has long represented the Wild West—the days of cowboys and sharp shooters romanticized by Hollywood.

In the late 1940s, John Ford started a rich tradition of filming here when he chose this as the location for several of his movies—many of which actually were shot on the Red Rock Ranch. The roster of stars that appeared in these pictures includes **John Wayne, Henry Fonda, Anthony Quinn, Ben Johnson,** and **Maureen O'Hara.**

The portrayal of the Wild West on the banks of the Colorado River actually wasn't stretching the truth too much. Butch Cassidy and his famous Wild Bunch often rode through the region, on the way to a heist or away from the law. The Red Cliffs Ranch itself was settled in the late 1800s, and has been in continuous operation as a working ranch ever since.

Valley onto Castle Valley Drive. This road is actually rather easy to miss, so keep a sharp eye out. A map on the right-hand side of Castle Valley Drive, just after the turn, confirms correct navigation.

Entering Castle Valley, it doesn't take long to understand what the fuss is about. To the immediate east of the road is the ultra-classic vista of **Castleton and Rectory towers.** Castleton is the thinnest, finger-like tower southern side of the formation; Rectory is the elongated one to the north. From the valley, elevation 4,500 feet, to the tops of the towers at roughly 6,660 feet, there is just more than 2,000 vertical feet of relief. Each of these towers has perfect, "splitter" cracks running more or less from the base to the summit—and are therefore home to several three- to four-pitch rock-climbing routes. The summit of each of these towers, as one could easily imagine from the ground, is nearly flat and quite interesting to sit atop.

Those who feel the desire can hike to the base of these towers. At a normal-to-brisk pace, the hike to the main saddle between Castleton and Rectory takes just more than an hour. Though most people hiking this path

FISCHER TOWERS: UTAH'S COOLEST MUDPILE

For those not in a rush, this is a small and worthwhile detour—requiring an extra 5.5 miles on UT 128 beyond the Castle Valley Drive. A total 21 miles east of US 191, look for a right-hand turnoff for **Fisher Tower Road.** Here stands a small (five-site) campground and 2.2-mile foot trail leading to the base of this formation.

Named for a resident miner in the 1880s, these unlikely fins looks more like upside-down mud drippings than stone towers hundreds of feet tall. Made of Moenkopi-topped Cutler sandstone, these have been naturally stuccoed with mud. These towers extend from a parent butte (to the north) that has gradually eroded away to expose three main fins and other smaller features. It is not diffi-cult to imagine that over time, the mesa will continue to erode, exposing more similar formations as it goes.

The towers have earned notoriety among rock climbers for their world-famous routes, particularly "Ancient Arts," which finishes atop the most pecu-liar-looking corkscrew summit. Anyone visiting the towers will understand immediately, though, that these routes have been deemed classic for their dis-tinctiveness—and most certainly not for high-quality rock climbing. This trail is on the eastern side of Castle Valley, so visitors to the towers will also get to enjoy views of **Castleton and Rectory** (or the "Priest" and "Nuns") towers, standing more than 2,000 feet above the valley floor.

Climbers Approaching Castleton Tower Zac Robinson

are climbers en route to the base of the climbs, the walk itself is quite worthy of the effort. The trail pleasantly gains elevation on gradually pitched switchbacks, picking its way up the talus and boulders beneath the formations.

The most easily reached trailhead for this hike is found by taking a left onto a dirt road roughly 4.7 miles south of UT 128. Drive along this to the obvious parking area and impromptu campground. The journey begins by scampering up a wash (requiring some agility and scrambling skills) until reaching a dirt road and plateau. As there are no signs, intuition will have to point the way to the main trail. As there are no trees on these talus slopes, route finding will not be difficult. A high-quality path leads all the

Fragile Desert Soil beneath Castleton

way to the towers—once found, this should lead directly to the saddle. Occasionally, a storm will wash away sections of this unvegetated trail—but, barring any severe weather, it should still be passable. Near the top, however, there is a particularly exposed section of trail. Though the walking there is not at all technical, a misstep could send the misstepper down a steep, steep mountain side. Keep the eyes forward and the balance in check.

Heading south in the Castle Valley, no turns are necessary to reach the **La Sal National Forest,** as Castleton Road turns naturally into the **La Sal Mountain Loop Road.** As the road winds up and out of Castle Valley, be sure to turn around and take in the increasingly panoramic views below. Though its towers are impressive when viewed from below, its integrated features are quite noteworthy from this retracted and elevated perspective. If you didn't have time to make the detour to the Fisher Towers, try to spot them now in the far backdrop (northeast side of the Castle Valley).

Several Forest Service dirt roads branch off of this main road, but the main route should be quite easy to follow. As it climbs, this road makes sweeping, dramatic turns along the foothills and plateaus of the dramatic **La Sal Mountains.** From north to south, the major peaks are as follows: **Grand View Mountain** (10,905 feet), **Horse Mountain** (11,150 feet), **Mount Waas** (12,331 feet), **Manns Peak** (12,273 feet), and **Haystack Mountain** (11,634 feet), with **Mount Tomasaki** (12,230 feet), rather hidden behind Haystack, to the east. As the road whirls around the skirts of this range, it might be difficult for the driver to keep track of each peak.

Almost the entirety of the La Sal Mountain Loop Road is in the **Manti–La Sal National Forest.** Because of this, camping is permitted at almost any reasonable spot away from the road. A few official campgrounds do exist—one at the end of Forest Service Road 063 (nearest the base of Mount Waas), and another at the end of Forest Service Road 076 (on Boren Mesa)—both short doglegs from the east of the La Sal Mountain Loop Road.

As the road makes one of its deepest bends toward the east and into a forested fold of mountainside, it passes over **Mill Creek** and **Mill Creek Canyon.** This canyon, though mostly hidden from view from the road above, is lined with more excellent sandstone walls—perfect for rock climbing. There is no published guidebook the area, so if you want to check it out, be prepared for adventure. Even attempting to descend into the canyon requires route finding, as rope ladders and the like are the way in.

MOAB CENTURY TOUR

If this route seems cool by car, imagine doing it on a bike! During the middle of September each year, a fully supported road ride takes place roughly along this drive. Distances range from 42 to 100 miles, and route options include out-and-back journeys on UT 128 along the Colorado River, or the burlier La Sal Mountain Loop Road in a counterclockwise direction. For those in town and wanting to participate, many bike shops offer high-end rental fleets right on Main Street in Moab.

Hang on for the turns and curves in the road ahead, and the road will eventually head back downhill to join Spanish Valley Drive. To reconnect with US 191, stay on this road as it heads north and roughly parallel to US 191. Take a left on County Road/ Old Airport Road for 0.6 miles, and then a right on US 191. Moab is just less than 8 miles north.

Though a short and unpopulated loop, this route toured many natural points of interest. The road passed through desert, slickrock, high alpine terrain, low, canyon floors, and by a giant laccolith—or igneous intrusion between two sedimentary strata—as in the La Sal Mountains. In addition, it passed one of Utah's most famous wineries, as well as one of its most photographed and filmed valleys. Once back in Moab, consider checking out chapters 6 and 8, for two other suggested drives that depart from that town.

IN THE AREA

Accommodations

Red Cliffs Lodge, Milepost 14 UT 128, Moab. Moderately priced. Cabins and guest rooms to rent overlooking the Colorado River; many other activities available by appointment. Call 435-259-2002. Web site: www.red cliffslodge.com.

Dining

Castle Creek Winery, Milepost 14 UT 128, Moab. Tours, wine tasting, and even dining available. Open year-round. Call 435-259-3332. Web site: www.castlecreekwinery.com.

Mill Creek Canyon in the High La Sal Foothills

Other Contacts

Colorado Riverway Recreation Area, Bureau of Land Management Office, 82 East Dogwood, Moab. Call 435-259-2100.

Fisher Towers Recreation Area, Bureau of Land Management Office, 82 East Dogwood, Moab. Call 435-259-2100.

Moab Century Tour. A three-day cycling event in the middle of September with rides and other events. Call 435-259-2698. Web site: www.skinny tireevents.com.

The Namesake "Needles"

CHAPTER

8

Moab to Canyonlands' Needles

Estimated length: 80 miles each way
Estimated time: Half a day

Getting There: Moab is within reasonable driving distance from three major cities: Salt Lake City, Denver, and Las Vegas. Salt Lake City is the closest, about five hours away. Those arriving from the east or west should simply come via I-70, turning south on US 191 at exit 182. Thirty-two miles later, this highway turns into Moab's Main Street.

To get to Moab from Salt Lake City, head south on I-15 for roughly 50 miles. Leave the interstate via exit 258 in Spanish Fork, heading southeast on US 6. This "shortcut" between I-15 and I-70 saves about an hour of driving and is actually quite scenic. Beware, though: the time-saving nature of this route makes it popular among every type of through-driver, from those in passenger cars, to RVs and tractor trailers. Be alert to other drivers' mistakes. US 6 dumps directly onto I-70. Head 23 miles east on this interstate, also using exit 182 for US 191.

Highlights: Moab is southern Utah's most famous tourist destination, surrounded by some of the world's most unique and dramatic geology. Not just pleasant to look at, the surrounding environs also serve as an excellent and accessible playground for all kinds of invigorating recreation, such as river rafting, mountain biking, rock climbing, and hiking.

This particular drive alone is scenic enough to offer a half day's entertainment just by sightseeing from pavement. On the way, though, it leads to numerous outdoorsy opportunities. En route, it passes arguably the world's best sandstone crack climbing area, **Indian Creek,** and ends up in the southeastern portion of **Canyonlands National Park**—perched above deep, steep canyons at the edge of the **Colorado River.** Here, trails lead down toward the rivers and smaller hikes afford views of ruins and geological spectacles.

Almost all features of this route are natural; don't expect towns, restaurants, or anything of that sort. The biggest man-made structure along the way is the **Canyonlands Visitor Center,** just inside the park.

This particular route begins in the outdoorsy tourist town of **Moab.** To those who haven't yet been, the environs surrounding Moab would seem to be all the same—desert, sandstone, and sagebrush—but in fact, this area has a stunning amount of diversity and natural depth. The groundcover morphs quickly and visibly with the slightest changes in elevation, nearness to a river, and hillside aspect. Each stratum of sandstone varies dramatically from the next in terms of color, texture, and thickness. And though the oranges, reds, and yellows of sandstone—the region's predominant rock type—dominate the foreground, the imposing **La Sal Mountains** provide a deep blue backdrop. Evergreen forests, sagebrush, cottonwood trees, and

---

INTO THE "WILD"

On this drive, don't expect to encounter any towns, restaurants, hotels or even reliable gas stations. With the exception of the **Hole N" the Rock** (a touristy gift shop) and **Needles Outpost** (with a small, seasonally open gas station just east of the Canyonlands National Park border), there are no on-route services beyond the southern end of Moab.

This journey is purely a sight-seeing, wilderness drive with opportunities to hike and explore the off-pavement environment. So before hitting the road, be sure to fill up the tank and pack a lunch, extra water, and sunscreen. A heads-up: cell phones typically don't work beyond US 191, and anyone breaking down is at the mercy of other passers-by. Those short on gas and unable to return all the way to Moab can head farther south along US 191 to **Monticello.** Though small, this town has multiple service stations and shops. Though out of the way, it is about 20 miles closer to Canyonlands than Moab—and sometimes 20 miles can make a difference.

cryptobiotic soil cover this impressive erupting from the skyline east of Moab.

This drive aims for the Wingate sandstone cliffs, fins, and towers sprouting out of broad canyons. Along the way, though, it must first ascend onto a high desert plateau as it passes the **La Sal Mountains.** It then plunges into the depths of **Indian Creek Canyon,** past sheer cliffs and steep talus piles, and arrives in the **Needles** portion of **Canyonlands National Park.**

To begin this route from Moab, head dead south on Main Street. As it leaves town, Main changes its name to US 191. The first 40 miles of the drive are the least scenic or exciting, though they do pass quickly and include a few points of interest. Look around and enjoy the stark natural surroundings and the subtle changes from one vista to the next. The only major north-south highway in the area, US 191 can sometimes get busy. Though this highway is predominantly two-lane, it does have semi-regular passing lanes—there's no need to overtake other vehicles impatiently. If stuck in a line of traffic, just wait a few miles and a chance to safely pass will come soon.

About 8 miles south of Moab, the **La Sal Mountain Loop** (Old Airport Road) departs from the highway to the east. Detailed in chapter 7, this commences a 60-mile loop that ascends the rising base of these impressive mountains, and ends by winding down through **Castleton Valley** and along the shores of the **Colorado River.**

Continuing south, the road dips steeply down and directly to **Hole N" the Rock,** about 16 miles south of town. This bizarre little shop is housed in none other than a rock—a massive outcrop of red sandstone. In addition to a kitschy retail store, there is also a petting zoo—open throughout the year. The total in-rock space measures 5,000 square feet and includes a gift shop and residence, adornments and furniture of which are carved directly into the rock. Over 12 years, this cavernous home was created by Albert Cristiansen, who removed 50,000 cubic feet from its parent rock. After his death in 1957, his wife, Gladys, transformed it into a public attraction, charging a small amount for tours of the residence.

Ten miles down the road (still on US 191) stands the obvious **Wilson Arch,** named for Joe Wilson—an early settler and resident in the area. This robust-looking entrada sandstone formation has a width of 91 feet and a height of 46 feet. As with other natural arches in the area, this was formed by the continuous freeze-thaw cycle in the desert whereby seeping water exploits cracks and weaknesses in rock as it freezes and expands. Pullouts exist on either side of the road and provide for safe viewing out of highway traffic.

About 14 miles farther along (roughly 40 miles from the southern end of Moab), keep a sharp eye out for the large yellow sandstone "submarine" approaching on your left. Officially labeled **Church Rock** on maps, this natural road marker signals the junction with UT 210. Slow down and prepare to turn right, or west, toward Indian Creek and Canyonlands National Park. The next few miles of driving uneventfully lead over high desert plains. Climbing slightly through sagebrush toward small, juniper-covered hills, be sure to keep a sharp eye for deer and free-range cattle. Especially near dawn and dusk, hundreds of deer can line the road on this stretch.

After miles of straights with almost no turns at all, the pavement suddenly plunges into the upper reaches of **Indian Creek Canyon** by way of very sharp, serpentine curves. Just before the road descends, the speed limit changes to 40 mph and curving-road advisory signs pop up. Beware—the bottommost S-curve is very severe and intimately adjacent to a very steep drop-off! Once at the canyon bottom, the road winds along a riverbed lined with cottonwood trees and growing sandstone cliffs. At the top of this creek basin, the canyon is narrow and cozy and the surrounding cliffs are still short.

Quickly after arriving at the canyon floor, **Newspaper Rock** pops up on the right hand side of the road. Pull into the parking lot to view an incredibly dense collection of petroglyphs. Carved onto a darkly varnished vertical slab, hundreds of figures have been protected from precipitation by a natural rock ceiling. The oldest of these rock etchings were created as long as 2,000 years ago by the Basketmaker, Pueblo, and Fremont peoples. Newspaper Rock was designated a Utah State Historical Monument in 1961. Today a protective fence stands around it, but all the rock art is clearly visible for no charge.

After leaving Newspaper Rock, UT 210 winds continues to wind gradually downhill through tight curves for another five minutes or so. As the creek eats deeper into the bedrock, the yellow, domed sandstone yields to incredibly sheer Wingate sandstone deeper down. As the miles tick by, the canyon broadens markedly into a gaping valley and the rock band's full height becomes exposed. These cliffs on either side of the road comprise one of the world's premier destinations for rock climbers interested in crack climbing. Keep an eye out for pullouts (especially on the northern

Newspaper Rock, at the Top of Indian Creek Canyon

Bridger Jack Mesa Viewed across the Creek at Beef Basin Road on a Cold Morning

side of the road); slow down, and on nice days, climbers will be scattered about the crags above.

The next few miles provide "nothing" but increasingly dramatic scenery as the basin expands even more. Multiple other drainages join together to form an ever-expanding basin surrounded by high talus piles crowned by sheer buttes. To the south, the **Bridger Jack Mesa** can be recognized as peninsula of rock that has been worn particularly thin. The pillars of Wingate that comprise it resemble a set of jagged teeth, and predict the future today's thicker mesas. It is possible to get a closer look at this formation and even camp near the base; take a left at the signed **Beef Basin Road** (8 miles after Newspaper Rock), noticeable by the outhouse at the turnoff. A very rugged dirt and rock road winds toward the formation and to primitive camping very near its base.

Continue westward as the road straightens somewhat and ascends slightly out of this valley. This is the final approach to **Canyonlands National Park.** Seventy-two miles outside of Moab, signs mark the park boundary. Just after crossing into Canyonlands, the turnout to **Needles Outpost** appears on the northern side of the road. This is a very small gas station/convenience store affair that is indeed an outpost. However remote the location, this joint still stocks a respectable spectrum of goods, from maps and camping supplies, to minimal groceries. There is also a grill-style simple restaurant and campground with showers and drinking water. Check ahead to be sure they actually are open for the season. Located so far from any town or highway, this shop has very limited supplies, so prices are elevated and availability is not guaranteed.

Forge on into Canyonlands. Just more than 3 miles into the park comes the fee station, where fees are collected and national parks passes accepted. A half a mile later you will find the **Needles Visitor Center.** Inside is a miniature museum giving an easily digested overview of the park (with a three-dimensional relief map of the park), as well as a handful of brochures, books, maps, and basic supplies for purchase. The rangers are quite willing to recommend hikes and drives, and talk about current trail conditions. From the here, one main, 10-mile-long paved road stretches in a southwesterly direction into the park. From this depart a few looping and dead-end side-spurs.

CANYONLANDS IN THIRDS

Though more or less one continuous landmass, Canyonlands consists of three geographically separate entities. The **Green River** flows into from its northwestern corner, and the **Colorado River** enters from the northeast. They join in the center of the park, forming a "Y" and effectively dividing the park into three regions. This southeastern portion, The Needles, is called such because of its predominant filament-esque rock formations. Though other parts of the park must be accessed by long, circumscribing drives, viewpoints here offer expansive views of the other regions of the park, particularly the northernmost region, **Island in the Sky.** When viewed from the Needles, the name's inspiration is obvious. To the west sits **The Maze,** the park's most wild portion, access to which is achieved primarily by high-clearance, four-wheel-drive vehicles and foot travel.

One of the first striking views is that of **Junction Butte** and the **Grand View Overlook,** to the northwest. Located across the Colorado, in the Island in the Sky portion of the park, these prominent buttes soar over the surrounding landscape.

Not even half a mile past the visitor center you will find the pullout for **Roadside Ruin** on the left side of the road. This stone and mud cylindrical grain silo was built under a rock overhang by ancestors of the Pueblo people between 1270 and 1295 A.D. The natural shelter, as well as the very dry climate, has allowed this structure to remain preserved almost perfectly for hundreds of years after its creation. A 0.3-mile-loop trail leads from the parking to the silo.

The next major roadside point of interest is the **Wooden Shoe.** This natural Cedar Mesa Sandstone arch was formed by rock of uneven strength. Millions of years ago, it was submerged under sea. Salt deposits from the ocean waters filled fissures in the rock that were later weathered away by above-ground erosive elements. Left behind is this wooden-clog-shaped arch. The pullout for this is 1.3 miles past the Roadside Ruin parking.

Just after the overlook, there is the option to turn left onto the Squaw Flat Spur Road, gateway to the **Squaw Flat Campground** and **Elephant Hill Road.** This scenic jeep and hiking loop winds down into some of the more colorful rock formations of the Needles District, including the **Devil's Kitchen.** The road itself was built by a bulldozer operator employed at a nearby cattle ranch in the 1940s. At that time, cattle ranching in the area was hot, and this road serviced small air strips used by planes delivering supplies to the ranches. Today, this dirt and 4x4 road is popular for jeepers. The first stretch is drivable in a "normal" car; but the second portion, a loop, is considered to be a challenging jeep road and is not passable in ordinary vehicles. Dogs are actually allowed on the first portion of this road—however caution is advised for pedestrians, as the road's twists and turns severely limit visibility.

Following the main road to its terminus, it bends almost due north and begins to wind more deeply into the desert sandstone strata. Look to the northeast to see a spectacular backdrop of the **La Sal Mountains**—the range you passed on your way out of Moab. Especially when this range is covered in snow, the La Sals and desert contrast quite powerfully. These deep blue slopes, often patched with white snowfields, form a striking backdrop for the red and yellow desert-scape, dotted with light green sage brush.

Ruins in Indian Creek's Beef Basin Mike Schenk

A small roundabout at the **Big Spring Canyon Overlook** clearly marks the terminus of the road (and the official end of this one-way drive). From here the **Confluence Overlook Trail** leads to a view of the confluence of the Green and Colorado Rivers. To reach the overlook, the hike is 5.2 miles each way. Though easy in terms of elevation loss (only descending about

The La Sal Mountains as a Backdrop over Canylonlands

200 vertical feet), it is long; be sure to bring plenty of water and sunscreen as it requires up to several hours to make the round-trip.

Worth seeing, these two powerful rivers flow together right in the center of Canyonlands—the Green from the northwest and the Colorado from the northeast. Some of the most important bodies of water in the history of the American West, these were famously navigated and meticulously charted by John Wesley Powell in 1869 and again in 1871. From the confluence, one can easily understand Green River's name; in contrast with the muddy, brown Colorado, this river appears to be nearly the color of emerald.

Just downstream of this junction is the famous **Cataract Canyon,** through which only knowledgeable rafters or guided clients may pass. This short and fierce stretch of river is the last, dreaded leg of a longer raft trip. Aptly named, the sheer walls of this canyon dramatically squeeze the Colorado River—now also with the waters of the Green—just before pouring into Lake Powell. This sudden narrowing of channel, as well very steep gradient, creates enormous and turbulent rapids that are highly unforgiving to all but the most experienced boaters.

At the end of the route, all that is left is to return to Moab, hoping the gas tank is full enough to make it. If in doubt, consider turning south on US 191 and driving to Monticello. Though not on the way for those wishing to return to I-70, its gas stations are 20 miles closer to Canyonlands than those in Moab.

IN THE AREA

Canyonlands National Park Office, 2282 SW Resource Boulevard, Moab. Call 435-719-2313. Web site: www.nps.gov/cany.

Canyonlands Needles Outpost, UT 211, Needles District, Canyonlands National Park, Moab. Open seasonally only; call for hours regarding specific services. Call 435-979-4007 . Web site: www.canyonlandsneedles outpost.com.

The Hole N" [sic] The Rock, 11037 South US 191, Moab. Open daily. Call 435-686-2250. Web site: www.theholeintherock.com.

The Kayenta Hills Start to Give Way to Towers David Sjöquist

CHAPTER

9

Monument Valley

Estimated length: 163 miles from Kayenta to Moab; 230 miles taking all suggested side trips

Estimated time: A full day with side trips; up to three days with stops for hikes, museums, and photographs

Getting There: Kayenta is located in the northeastern corner of Arizona, on US 160, about 100 miles east of Page and about 40 miles west of the US 160/US 191 Junction.

Highlights: Though this drive actually starts in Arizona—oops!—it would be a scam to write a book on Utah's scenic drives and not include **Monument Valley,** one of the most photographed and iconic routes in the American Southwest. The nearby, under-the-radar scenery between **Mexican Hat** and **Blanding** includes ruins, museums, state parks, and national monuments of great scenery and complementary solitude.

This route begins **Kayenta,** in the **Navajo Nation,** among tombstone-like towers 1000 feet tall, and continues across desert planes, over the **San Jan River** to the many state parks and natural attractions between Mexican Hat and Blanding—**Goosenecks State Park, Natural Bridges National Monument, Valley of the Gods**—and eventually into **Moab.** The drive through the valley itself is almost completely devoid of human infrastructure, save for the road itself.

Leaving Kayenta, Arizona   David Sjöquist

Kayenta, Arizona (or **Tó Dinéeshzhee,** in Navajo) is part of the **Navajo Nation (Diné Bikéyah),** a 17-million acre state occupying the northeastern part of Arizona, southeastern portion of Utah and northwestern corner of New Mexico. Kayenta itself was settled in 1910 as **Oljeto,** first as a trading post, later fueled by uranium mining activities. Today it is a small and dusty town that offers a spread of basic services for tourists—from an auto mechanic to restaurants and hotels. Among a majority of national fast food chains and motels, a few local restaurants (and one hotel, north of town) present an alternative choice.

For Southwestern/Mexican cuisine with a Navajo influence, stop by the **Amigo Café.** A standard fare of combination platters, burritos, and enchiladas is appended by the very popular Navajo Taco. This restaurant is visibly the most popular in town among locals and tourists. Make sure to have cash in the wallet as it is the only accepted method of payment. You won't need much, though, as a small amount of money purchases large portions. The **Blue Coffee Pot** offers a menu heavier in Navajo cuisine. A tip: order within the Navajo selections, as they are the kitchen's strength. Frybread, Navajo tacos, and stews beat the alternative fare.

National, mid-level and budget chains supply the guest beds in town. While these ensure a familiar style and level of quality, **The View Hotel** in Monument Valley Tribal Park (described below) offers a vastly more local flavor. This hotel stands just south of the Utah-Arizona border, so not too far from town.

To embark on the drive, take US 163 north directly out of Kayenta center. Climbing slightly on the way out of town, it seems unlikely this should be the correct route. Suddenly, though, a few turns and dips reveal the immediately familiar and awe-inspiring view of Monument Valley. One of the most barren, yet striking, pieces of desert stretches into view. In all the valley, it seems the only man-made intrusion is the highway. To either side rise pillars of sheer Wingate sandstone, atop islands of red talus. Everyone has seen images of

KAYENTA FOURTH OF JULY RODEO

The Kayenta Fourth of July Rodeo takes place every year July 1–4. Much more than a quick-and-dirty rodeo, this four-day, town-wide celebration includes an all-Indian rodeo, youth rodeo, team roping, barrel racing, bull riding, a pancake breakfast, and massive fireworks finale on the Fourth of July, called the "Best of the Best." As of 2010, this rodeo has won many awards, including seven "Rodeo of the Year" and two "Outstanding Rodeo" designations.

this valley, though offhand, it's hard to know where. The first film shot here, *The Vanishing American,* in 1924, began a list of 81 titles (as of 2010). Among the most famous are the 1939 *Stagecoach* with John Wayne, *Easy Rider* (1969), and *Forrest Gump* (1994).

About 21 miles north of Kayenta (halfway to **Mexican Hat**), US 163 crosses into Utah. Just after this, look sharp for a right-hand turn to **Monument Valley Tribal Park.** Though this famous valley can be viewed fairly adequately from US 163, this park offers a relaxed and picturesque detour. Open 365 days a year, barring extreme weather, the route adds a few hours to the trip. In the park wait a visitor center, scenic drive, and campground, just as there would be in a National Park Service–operated affair. Stop in at the visitor center to learn a bit about the religious and cultural significance of this valley. Additionally, there is a gift shop and **The View Hotel**

THE MONUMENT VALLEY FILM FESTIVAL

Founded in 2007, this takes place in Kayenta around the third weekend of September each year. After the completion of their debut full-length film, *Mile Post 398*, Shonie and Andee De La Rosa established this—the Navajo Nation's only film festival at present. All films are written, directed, or produced by Native Americans. Sheephead Film's 2007 *Mile Post* tells the story of a Navajo man facing drug and alcohol problems, his decision to quit and resulting struggle to make amends with family.

(see sidebar). The tribe collects an entrance fee (National Parks and Golden Eagle passes not accepted), and hours are roughly confined to daylight hours.

The only hiking trail in the park not requiring of an authorized Navajo guide, the **Wildcat Trail,** departs from the campground just inside the park boundary. A 3.2-mile loop with a bit of elevation change, this requires some effort but is still moderate in distance and elevation gain. The journey begins with a descent into the valley and proceeds to make a loop around the **Left Mitten Tower.** (From this trail, its partner, **Right Mitten,** is also readily visible.) Navajo take this path in a clockwise direction just as they would walk in a *hogan*—the traditional, rounded permanent home sacred in the Navajo religion.

Those sticking to the road will embark upon a 17-mile, high-grade dirt road loop among the towers. Self-guided tours are obviously free beyond the entrance fee, but paid tours are also an option, costing roughly $50 to $100 per person, depending the route. Various portions of the park, however, like **Hunts Mesa** and **Mystery Valley,** may only accessed by way of guided tours. If interested, inquire at the visitor center or call the park.

Pavement extends east beyond the visitor center for another few miles before turning to dirt. Around this transition, the **Left and Right Mitten towers,** whose shape should make their identity obvious, rise out of the desert floor to the left (north). More than a dozen other named buttes will be visible on this drive, so those interested in pinning down their identities should pick up a brochure labeling them.

Continuing along, the road dips subtly toward the south and heads into the main thicket of towers. Just after passing **Elephant Butte** on the left, the road enters its loop portion, encircling **Rain God Mesa.** Around this formation stand several others in the foreground, their relative position ever changing when passed on this curving road. Don't expect to see much wildlife—populations of all kinds are tempered by the extremely dry climate—but do notice the purple sage and junipers that contrast strikingly with the deep red desert soil and rock. Soon the loop ends and dumps onto the same road driven into the park to reach it. Do not attempt to depart onto side roads; these are closed to parties unaccompanied by an authorized Navajo guide.

Leaving the park, head back to US 163, turn right, and continue farther north into Utah. Up next is perhaps the nation's most famous single stretch of road—and perhaps one of the few instances in which a highway itself is an exciting attraction. Finally, the towers of Monument Valley fade in the rear view mirror. Crossing over the **San Juan River,** the pavement comes into **Mexican Hat,** about 20 miles beyond the park turnoff. For the passerby, this tiny town serves as little more than a junction on the way to and from the surrounding state, national, and tribal parks. Named for a rock formation, a 60-footwide capstone disk atop a much smaller cylinder—the "Mexican Hat" properly resembles an upside-down sombrero. As of the 2000 Census, this town had a whopping 88 residents among 22 families.

THE VIEW HOTEL

Located adjacent to the tribal park's visitor center, this hotel warrants a stop, even for those not requiring a place to bed down. The young owner, a sixth-generation Navajo in a family with long-standing stewardship of the Southwest Native American arts, has saturated the entire facility with this legacy. The hotel showcases contemporary Native American arts and jewelry as well as recent and ancient historical artifacts from the valley. On display are photographs from Hollywood pictures filmed in Monument Valley and tribal memorabilia.

The on-premises **Trading Post** offers wares for sale, the showpiece of which is its Navajo woven rug collection, the largest within 250 miles. Additionally, there are hats, pottery, jewelry, and other art pieces. This is no rinky-dink souvenir shop; most of the goods for purchase would sooner be called art than trinkets (though less-serious crafts are sold here, too).

That said, the View Hotel also represents the only in-park accommodations, and likely the finest for miles and miles around. Opened December 2008, this 95-room, eco friendly establishment was built by, and is today still owned and operated by, Navajo. Be sure to make reservations (particularly during the summer), as hotel has no peer in the area. No alcohol is served here as it is forbidden by the Navajo Nation.

Mexican Hat does, however, offer some amenities. Travelers wanting to spend more than a day casing the area can stop here and rest up at the **San Juan Inn and Café.** Possibly one of the more unique lodging options around, this inn rents guest rooms in the hotel, as well as yurts and…trailers? Described as "Unique American Lodging," these trailers sit on raised, cinderblock foundations, have aluminum sides, and are bona fide trailers! But the San Juan's yurts could possibly be the coolest options. From the outside, they appear primitive, but these private structures actually have modernized, simple interiors. The hotel itself abuts directly to the cliff edge above the San Jan, affording exposed views of the water below. As with any desirable lodging options in the area, be sure to call for reservations in peak season.

In addition to lodging, the San Jan Inn runs a restaurant and **Trading Post** (read: convenience store) on the premises. The **Old Bridge Grille,** open daily and year-round, serves an American menu supplemented with a few classic Navajo dishes. Outside of the Navajo Nation, this establishment can legally serve liquor, wine, and beer. This is an ultra-casual Western joint with pool tables. The San Juan Trading Post carries basic wares: beer, snacks, maps, and books. Its main focus, however, is on Navajo and other Native American arts: rugs, pottery, clothing, and jewelry.

NAVAJO NATION

As of the 2000 Census, there were roughly 300,000 Navajo living in the United States, and about 60 percent of these resided in the Navajo Nation itself. This enormous tract of land is, at 17 million acres, larger than Lithuania. In the 1920s oil deposits were discovered on Navajo land, and in 1923 an official tribal government was formed to help organize the leasing of this land by American oil corporations. Since then, this administration has grown to cover the general needs of the nation and people, and today's Navajo Nation Council has 88 delegates representing 110 chapters.

Leaving Mexican Hat, prepare for a desolate stretch of road! Though the scenery is beautiful, almost no infrastructure interferes with viewing it in near perfect solitude. The next destination is **Goosenecks State Park,** which you will reach in about 10 minutes by driving north of town on US 163 East and taking a quick left on UT 261/316. (The state park sits at the terminus of 316.) From the center of town, the total drive is just a five or ten minutes. This state park overlooks a severely curvy section the San Jan River. About 1,500 feet lower than the desert surface, more than 6 miles of

Crossing the San Juan River   David Sjöquist

river wind in a linear distance of just 1.5 miles. Over the course of a 300-million-year erosion process, it has gently revealed a fine brown, gold, tan, and orange rainbow of horizontal rock strata. These extremely long river loops create a maze of interlocking, fingerlike peninsulas. In the distance are **Monument Valley** and **Alhambra Rock,** an exposed volcanic intrusion of dark rock. The park functions primarily as an overlook point, with very simple infrastructure, and the principle activities here are gawking and photography. No drinking water is available here, but picnic tables are provided and primitive camping permitted; no entrance fee required.

After Goosenecks, the next stop is **Valley of the Gods,** a very Monument Valley-esque area. A collection of scattered rock pinnacles in a flat valley, this is home to smaller, but often more peculiar, spires than those of

Hills outside of Mexican Hat

Monument. Valley of the Gods can either be seen by driving through it (another 17-mile dirt loop, but this time free of charge) or bypassing it and viewing it from the top of **Cedar Mesa** after ascending the **Moki Dugway.** To tour this valley, take UT 316/261 back south to US 163. Take a left to head northeast on this highway, watching for a dirt road, UT 242, on the left (in about 4 miles). Head north into the Valley of the Gods here. This road follows **Lime Creek,** a wash usually only flowing in spring, and then snakes around the rock formations in a westerly direction before popping out on UT 261, just south of the Moki Dugway. Though a short loop, it wouldn't hurt to carry a map.

The Valley of the Gods is, but for the road, entirely undeveloped and offers much more solitude than the already quiet Monument Valley. Entirely on BLM land, all camping here is free and unregulated. Though no official trails exist, overland, desert-floor wandering is permitted—best done in dry washes, so as to avoid walking on and destroying cryptobiotic soil. Bring a detailed topographic map along to assist with identifying formations whose names range from **Rooster Butter,** to **Lady in the Bathtub, Rudolph,** and **Santa Claus.**

Among 360,000 acres in the **Cedar Mesa Cultural and Recreational Management Area, Valley of the Gods Bed & Breakfast** is the only house! Solar- and wind-powered and fire-heated, this self-sustaining building requires no outside power source, and its owners must manually haul water from Mexican Hat. Originally built in 1933, the structure has seen serious restorative renovations, but it only became a B&B years later in 1993. The unique rooms are each lit by oil lamp in the evening, and beds and walls are covered with Southwestern and Navajo rugs. Construction is of two-foot-thick sandstone and limestone block walls, with sturdy beam work crafted of recycled timbers. Breakfast here is held outside almost always, but inside when the weather fails to cooperate. The bed and breakfast is located just east of UT 261 on Valley of the Gods Road/UT 242. To preserve the solitude for other guests, pets and children must be accepted by prior approval. Call ahead for reservations, as rooms are very limited.

Moki Dugway is a 3-mile section of UT 261 just north of the Valley of the Gods B&B. A stretch of road that would never be cut in modern times, this switch-backing dirt road was built for uranium mining trucks in 1958. Sustained sections of 10 percent grade aggressively ascend the steep side of **Cedar Mesa,** climbing more than 1,100 feet to an elevation of 6,424 feet. Many signs announce the meanders as they approach, warning off heavy vehicles. Soon the two-lane highway shrinks to a one-lane paved street, and eventually a graded dirt road. As traffic is still two-way, the grade so steep, and the switchbacks so dramatic, the recommended speed is just 5 miles per hour. At the top of the ascent, be sure to look back south over Valley of the Gods and Monument Valley for a scene of almost completely undisturbed landscape.

Continuing north on UT 261, look for signed turnoff (very soon after Moki Dugway) to **Mulley Point.** Five dirt road miles lead back to the edge of Cedar Mesa once again. This truly spectacular cliff-edge viewpoint offers a 360-degree panorama that summarizes the entire drive so far. From atop strewn boulders and fractured sandstone, all of the "valleys," the Goosenecks, Alhambra Rock, and Navajo Mountain are all visible to the south and west. Because of the dramatic relief and steepness of canyon walls, the San Jan River remains out of sight. West of Muley Point, the San Jan River flows into the Glen Canyon National Recreation Area, as the easternmost tributary for **Lake Powell.**

OK, hikers—here's another chance to get out and super-stretch the legs. The **Grand Gulch National Primitive Area** is a 35-mile-long gulch that, rarely visited, is lined with undisturbed petroglyphs and pictograms. Leaving on foot from Kane Springs actually requires quite a hike just to access the gulch: up to five hours each direction. The BLM requires visitors to preregister, which is a good idea anyway when visiting such a wilderness area. As this is a delicate archaeological area, all pets must be on leashes at all times. For directions and permits, visit the **Kane Gulch Ranger Station,** about 4 miles south of UT 95 on UT 261.

Finish out UT 261 by heading north to **UT 95.** A very quick jaunt to the west and then a turn to the north on UT 275 leads directly to **Natural Bridges National Monument.** With just three arches in the area, this is like a small, concise version of Arches National Park. In a place with just a few arches, visitors can pay more attention to the area's human history than to holes in rock. Three primary hiking trails, one each for **Sipapu Bridge, Kachina Bridge,** and **Owachomo Bridge** lead away from the pavement and tour the backcountry. Petroglyphs, pictograms, ruins, and other artifacts have been left behind here by the Hisatsinom (a people often included under the "Ancestral Puebloan" blanket term), ancestors to the Hopi. These people occupied the area during three distinct and separate eras, each time returning with much different tools and agricultural technology. The earliest occupation was around 200 A.D., and the latest ending around 1270.

One of the last stops on the way into Blanding via UT 95 is **Mule Canyon Ruin** on the north side of the road, roughly 12 miles east of UT 261 (and 19 west of Blanding). This complex of above- and below-ground structures dates back to around 750 A.D. Archaeological evidence suggests that these people were predominantly Mesa Verde Anasazi, with some Kayenta Anasazi influences. The ruins consist of 12 rooms that were used as living and storage space, a kiva (or center of ceremonies), and a tower. This is located about 0.25 miles north from the UT 95 turnoff; park and walk into the canyon to the west. Operated by the BLM, this area has no entrance fee.

Nobody would guess it at first glance, but Blanding is actually home a few museums. **The Edge of the Cedars State Park Museum,** right in town, is situated around an actual village that was built and inhabited by the ancestral Puebloan from roughly 825 to 1125 A.D. Visitors, if able, can climb down a ladder into one of the original structures. A walking trail tours the ruin, and the grounds have been landscaped with native plant species. The museum itself hosts traveling and permanent exhibits pertaining to art and human history, and it showcases historical artifacts such as pottery and historical and contemporary photography. This museum holds the region's largest collection of Anasazi pottery. Every May (on the first Saturday of the month), an annual Indian Art Market takes place here

Also in the town of Blanding is the **Dinosaur Museum.** Paleontology is an especially active science in the Four Corners region as the lands have been richly littered in fossils. This museum draws from the richness of locally active scientists and their work, but also includes dinosaur remains from around the world. Here, history is explained through plate tectonics, showing where the different continents of land have been located at different points throughout history, how this dictated where certain fossils are found today, and why different mountain ranges formed. A genuine meteorite is even on display in the museum.

After Blanding, the drive is pretty much up. The most direct route back to Kayenta is via US 191 and US 160. To get to Moab, head north on US 191 about 75 miles; I-70 is just another 30 miles north of Moab. If staying in Moab, check out the three Moab-based drives in chapters 6, 7, and 8 of this book.

IN THE AREA

Accommodations

Valley of the Gods Bed & Breakfast, on Valley of the Gods Road (UT 242), just east of UT 261. On the pricey side. Eco-friendly, off-the-grid, historic home. Call Gary or Claire Dorgan, 970 -749-1164. Web site: www .valleyofthegods.cjb.net.

The View Hotel, Monument Valley Navajo Tribal Park, Monument Valley. Somewhat pricey. Modern hotel embellished with Navajo culture. Call 435-727-3470. Web site: www.monumentvalleyview.com.

Attractions and Recreation

The Dinosaur Museum, 754 South 200 West, Blanding. Open April 15–October 15 only, Mon–Sat. Very modest admission fee. Call 435-678-3454. Web site: www.dinosaur-museum.org.

Edge of the Cedars State Park Museum, 660 West 400 North, Blanding. Open year-round. Small admission fee. Call 435-678-2238. Web site: www.stateparks.utah.gov/parks/edge-of-the-cedars.

Goosenecks State Park, northwest of Mexican Hat, at the terminus of US 316, Mexican Hat. Call 435-678-2238. Web site: www.stateparks.utah.gov/parks/goosenecks.

Grand Gulch National Primitive Area, Kane Gulch Ranger Station, about 4 miles south of UT 95 on UT 261. Permits required to enter; contact the BLM for directions. Call 801-587-1500.

Monument Valley Navajo Tribal Park and Visitor Center, Monument Valley. Fee required to enter; national parks passes not accepted. Open year-round; visitor center closed Thanksgiving and Christmas. Scenic Drive open 365 days a year, barring severe weather. Call 435-727-5874 or 435-727-5875. Web site: www.navajonationparks.org.

Mule Canyon Ruin, Bureau of Land Management, San Juan Field Office, 435 North Main, Monticello. Call 435-587-2144.

Classic Monument Valley View in Rear View Mirror David Sjöquist

Natural Bridges National Monument, HC-60 Box 1, Lake Powell. Call 435-692-1234. $3 individual entry, $6 per vehicle; all federal lands passes accepted. Web site: www.nps.gov/nabr.

Dining

Amigo Café, US 163, Kayenta, AZ. Affordable. Open daily for breakfast, lunch, and dinner. Mexican and Navajo cuisine. Cash only. Call 928-697-8448.

Blue Coffee Pot, US 163 and US 160, Kayenta, AZ. Affordable. Open Monday through Saturday for breakfast, lunch, and dinner. Navajo cuisine. Call 928-697-3396.

Old Bridge Grill, see San Juan Inn and Café, below. Very affordable. American and Navajo cuisine; casual. Open daily, year-round.

San Juan Inn and Café (Old Bridge Grille, and **Trading Post),** US 163 and the San Juan River, Mexican Hat. Moderately priced. Hotel rooms, yurts, and trailers for rent; general store and restaurant onsite. Call 1-800-447-2022. Web site: www.sanjuaninn.net.

Other Contacts

Kane Gulch Ranger Station, about 4 miles south of UT 95 on UT 261. Call 435-587-1500.

Kayenta Fourth of July Rodeo. Annual, four-day celebration in Kayenta. Call Chris Claw, 928-675-1035. Web site: www.kayentarodeo.com.

Monument Valley Film Festival. Call Shonie De La Rosa, 928-429-0671. Web site: www.monumentvalleyfilmfest.com.

San Juan Trading Post, see San Juan Inn and Café, above. Open daily, year-round.

Cirque in the Henry Mountains

CHAPTER

10

Henry Mountains and the Burr Trail

Estimated length: 170 miles without side trips
Estimated time: Nearly a full day with sightseeing stops

Getting There: This suggested route begins in the town of Green River and ends in Boulder (Utah). Green River is located just north of Moab, right on I-70 between exits 160 and 164. From Salt Lake City (or any northern point of origin), the best approach is via I-15 South, and then US 6 and 191 South. This diagonal cut-across saves more than a hundred miles—though, a high mountain pass, should be taken in good weather only. From the east or west, simply arrive by way of I-70. Almost no one arrives from the south as the towns in that direction are small and distant and Lake Powell essentially barricades any potential approach paths.

Highlights: The southern **San Rafael Swell** and **Henry Mountains** are in a wonderfully remote region of south-central Utah. Neither near Moab nor Zion Nation Park, they remain largely vacant, with miles of uncluttered scenery and rare solitude. A geologically spectacular journey, this route heads south along the eastern front of the Henry Mountains to the northern shore of **Lake Powell**, then loops west on the historic **Burr Trail**, and traverses the **Grand Staircase-Escalante Region**.

Note: the Burr Trail, a key component of this drive (from Bullfrog to Boulder), contains a very steep, dirt section and should be taken in good weather only. Though the road is primarily chipsealed, a 17-mile stretch (though **Capitol Reef National Park**) remains grated dirt. It is not recommended for RVs or vehicles towing trailers, regardless of conditions; though the dirt surface is typically fine, the sharp twists and steep grade render it inappropriate for heavy, long vehicles.

Green River exemplifies the classic interstate town. Big truck stops, dreary landscape, desolate feeling. This comes as a surprising fact as Green River must be passed by those traveling en route from Salt Lake City to Moab, Zion to Denver, and Grand Staircase-Escalante National Monument to Arches and Canyonlands national parks. The impersonal façade, though, is a bit misleading, as this town sits at the gateway to much of Utah's best canoeing, kayaking, and whitewater rafting (on the Green River).

Because of its location on the Green River, and because of this river's significance in the history of the West, the town has erected here **The John Wesley Powell River History Museum.** In a small, unimpressive town, this museum is a refreshing and educational surprise. Though it passes quietly through town, the Green River holds value in United States history. This museum communicates this, as well as the stories of the people who explored its waters.

For a quick meal to start the trip, **Ray's Tavern** is the best bet for those seeking something other than fast food or a truck stop diner. Reputed among tourists statewide, this place serves a good burger and cold beer. It can draw quite a crowd for its petite structure, but the tavern offers bar seating in addition to tables. In this small joint with a rustic atmosphere, it's OK to come dressed in Wranglers or in bike shorts. Beware: As with what seems like the entirety of rural Utah, this place stays closed on Sundays.

To begin the drive, head west just briefly on I-70, watching the miles count down to exit 149. Take this and drive south (the only option) into the San Rafael Desert. Continue along UT 24 about 23 miles until you see a sign to the right for Goblin Valley Road and **Goblin Valley State Park.** Make this turn to the west, following signs (and two more left-hand turns) to head south on UT 303 and into the park. Though it might sound complicated, this detour requires only 10 to 15 minutes and is well signed.

Aptly named, this valley appears to be filled with sandstone monsters—quite a sci-fi look with thousands of rounded statuettes standing shoulder to shoulder. For a $7 day-use fee, visitors have access to a visitor center and some of the most peculiar rock formations in existence. Though these hoodoos are not necessarily unique to this valley, they are neverthe-less in extremely high concentration here. These mushroom-shaped figures are formed when an erosion-resistant rock layer sits atop a weaker stratum. As time passes, the bottoms layer erodes more quickly, leaving a rounded capstone resting atop a thinner base.

As with any natural phenomenon of such distinctive character, this was a compelling place for indigenous tribes as well as early cowboys and pio-neers. The Ute, Paiute, and Fremont left significant rock art in the form of pictographs and petryglyphs. Cowboys and pioneers used it as a place of reference for travel, and occasionally as a hideout. In 1964 this was desig-nated a state park to help protect it from vandalism. Visitors today are welcome to hike among the goblins.

Return to UT 24 and head south to **Hanksville.** As the road approach-

MAJOR JOHN WESLEY POWELL

Powell was one of the earliest and most influential explorers of the Intermoun-tain West. A true renaissance man, he was considered an expert in botany, eth-nology, history, surveying, geology, and more. Born in 1834 in Mount Morris, New York, John Wesley Powell was a largely self-educated man who fought in the Civil War for the Union Army. During the war, John lost his arm in battle, making him an unlikely candidate for what would later become a career in rugged river explorations.

Prior to the late 1860s, the interior of the West was all but completely untamed and uncharted. Though occasional trappers and traders had penetrat-ed the region, there were no reliable maps or records of the area, and the U.S. Government wished to know more. In 1869 (and again in 1871) Powell tra-versed the entirety the Green and Colorado Rivers from Green River Wyoming to the Gulf of California. Along the way, he made scrupulous notes and drawings of everything pertaining to weather, wildlife, geology, and native peoples.

The immensity and strangeness of his expedition required a braveness that likely cannot be understood day. Not only were the lands, wildlife, and indige-nous people completely unknown, Powell's mode of travel was completely new and quite dangerous. White-water boating wasn't heard of. Yet, some of the biggest drops in North America are found along his route. Powell described this strange phenomenon in his journals, detailing their strategies for passage, as well as accidents and equipment losses incurred on the way. Not without accidents and losses, his crews were mainly successful—led by their one-armed captain.

A Better-than-Normal Dirt Road in the Henries

es this small town, the **Henry Mountains** will become unmistakably visible to the South. As with the La Sal Mountains near Moab and the Abajo Range, these isolated peaks rise high above their surroundings, standing alone. Roughly 25 million years old, each range was formed by magma forcing itself high into earth's crust. The magma cooled gradually while still under ground. Over time, the sedimentary layers atop these intrusions eroded away, exposing rock classified as porphyry. This volcanic rock has large crystals formed by the slow, subterranean cooling of the mountain-forming magma.

Because of its height, this range visibly creates its own weather system, collecting clouds that then drain their precipitation onto the slopes below. Even if the weather in Hanksville is sunny and warm, storms can be unleashing on the Henrys. Because of this heightened rainfall, be exceedingly cautious if entering the range on dirt roads. The nature of this soil causes it to clump and gather on wheels while wet.

Before heading into this range, be absolutely sure to refill the car with all the necessary fluids, the bottles with water, and the bellies with food.

Blondie's Eatery & Gift Shop is a diner-type affair where you can order at the counter and don't have to mess with table service. On the menu: burgers, chicken, salads, sandwiches, and espresso. A big stone sign out front makes it hard to miss. Call ahead for takeout from **Bull Mountain Pizza Deli.** Enjoy food outdoors or take it to the campsite or hotel; this restaurant has no dining room.

Curious about paleontology? Just dying to know more about the late Jurassic Period? On the way out of town, hit the the **Hanksville-Burpee Quarry,** a 150- to 600-foot dig into the Morrison Formation. Part of the late Jurassic Period and roughly 145–150 million years old, this site has been long-known to locals for its richness in fossils. But it wasn't until 2007 that the Rockford, Illinois, Burpee Museum of Natural History applied for digging permission and learned of the region's true wealth. Approved, they

---

BUTCH CASSIDY'S WATERING HOLE

Hanksville, a tiny mining and agricultural town, is the last stop before the Henry Mountains and the general wilds beyond. It also serves as a filling station for passers-by as it sits directly on the route to the Henry Mountains, Goblin Valley State Park, Capitol Reef National Park, the Grand Staircase-Escalante National Monument, the takeout for Cataract Canyon on the Colorado River, and the northern tip of Lake Powell,

Actually, in times past, it served the same purpose as a filling station. During their outlaw tour of the Old West, the legendary Butch Cassidy (Robert LeRoy Parker) and the Sundance Kid (Henry Longbaugh) regularly hid out in the Robber's Roost, just to the southeast of Hanksville—beginning in the 1880s. They made sure to establish friendly connections with the people of Hanksville so that they could be resupplied with beef, food, and other wares without having to surface in the face of the law.

With extremely rugged, maze-like terrain and natural lookout towers, this "roost" provided the gang shelter from the law, as they could easily see parties approaching and defend their post. Here they stashed weapons, ammunition, and other restorative supplies. Today **Robbers Roost Canyon,** a tributary canyon to the **Dirty Devil River,** is named after this hideout. Additionally, the original Wild Bunch corral, various carvings, and a stone chimney still stand. Here there have been established many moderate-to-technical slot canyon routes. Any visitor to the Robber's Roost Canyon today will understand how it was never penetrated by the law.

LAKE POWELL'S BATHTUB RING

Lake Powell is a massive reservoir, and the Glen Canyon Dam one of man's greatest architectural accomplishments. Built over the course of 10 years—from 1956 to 1966—this 710-foot-tall dam has the capacity to store 30 cubic kilometers of water between the walls of the Glen Canyon. Today it is responsible for up to 13 percent of power production in Utah and 6 percent in Arizona, as well as the supply of more than 10 million cubic kilometers of water to Arizona, California, New Mexico, and Nevada.

However, not all reports on this reservoir are positive. In addition to the massive environmental disruption it has caused upstream and downstream of the dam, the reservoir itself has faced serious logistical problems.

In 1983, following a very snowy winter, a warm and wet spring fell swiftly upon the area surrounding Glen Canyon. The above-average rainfall and rapid temperature increase conspired to pump the Colorado River full of water and inundate the reservoir with a sustained river flow of more than 100,000 cubic feet per second—a veritable deluge. By necessity, the spillways were very cautiously opened for the first time in the dam's employment and, due to the incredible flow of water, were indeed damaged. Massive chunks of sandstone were ripped from the spillways, and there was even concern that the entire dam might fail—but strategic operation kept damage in check. During this time, however, the water level rose to within six feet of the dam's crest.

Fast forward almost 20 years to one of the most severe and prolonged droughts in the history of the southwestern United States. In 1999 the water levels were relatively high, stable, and under control. Enter a severe, multi-year drought. Over the next few years, the water began to recede dramatically. As this happened, white rock was exposed on the sheer, lake-side canyon walls, and what was formerly reservoir returned to river at the inlet tips of lake. New miles of river rafting formed on the Colorado River beneath Cataract Canyon. Reservoir-bed sediment was exposed and turned suddenly into rapid-causing, rock-like boulders. By 2005, water levels dropped to record lows—at which time, the reservoir was filled to only one-third capacity. As of 2010, the water level is on the rise again, though at 63 percent full, the waterline is still almost 70 feet below maximum.

began excavating in the summer of 2008. Only three weeks into the dig, the site yielded a possible stegosaurus, two unknown carnivorous dinosaur, and multiple sauropods. The Burpee Museum provides a public tour detailing and demystifying the site's science.

Now begins the wilderness! Head south from town on UT 95. Though our drive turns west onto UT 276 about 30 minutes south of town, UT 95 leads directly to the northern tip of Lake Powell. Those interested may take this detour, but the Hite Marina, located there, is now out of commission. Hite was once the northernmost marina in the lake. It sits exactly where the Colorado River joins the lake and slows down—immediately after one of the most turbulent stretches of whitewater in the United States: Cataract Canyon. In the last few decades, however, Lake Powell has been evaporating at a rate faster than its tributaries can fill it, and the water level has fallen more than 120 feet. Beneath the natural red sandstone, a giant "bathtub ring" of white rock encircles the entire lake—readily visible in this area. As the waterline dropped, so too the shoreline receded. Hite tried to stay in business by extending its boat ramp several times. Now defunct, the Hite Marina looks much more like a landing strip for jet planes than for boats.

Those not taking the Lake Powell detour should look for a right-hand turn onto UT 276, about 30 minutes south from Hanksville (and long before Hite Marina). From the valley (elevation 4,300 feet), consider the Henry Mountains, try to guess which peak is Mount Ellen, the highest in the range at 11,506 feet.

The Henrys can be accessed by a number of dirt roads heading up through washes and foothills, and eventually to the mountain saddles. This range is altogether wild, and, aside from the roads, there have been no "improvements" made to them—not even trails. However, as almost nobody drives here, any kind of hiking desired can be accomplished right along the roads. As of my visit, this network of roads was altogether unsigned. Those with a specific peak in mind should take a good guess at which road will lead to it, and follow the eye. In June 2003, a 34,000-acre wildfire consumed much of the veg-

Victim of Massive Wildfire

View of the Horn, Between Mt. Ellen and Mt. Penellen

etation here—the scars from this are quite visible today. This blaze rendered high visibility across much of the range.

Many, many dozens of miles of dirt road exist in this weather-battered range, and rain and snow can wash out roads and expose oil-pan-eating rocks much more quickly than these mountain lanes can be repaired. If entering this range, consider it a wise choice to do so in good weather only, with high clearance, and four-wheel drive. The clays in the soil here make the mud extremely tacky and can create major problems for a vehicle.

The range spans an elevation rise of roughly 1.5 vertical miles with correspondingly diverse fauna. Lower yucca, cactus, and sage—characteristic of warm, dry deserts—blend into rabbit brush, scrub oak, juniper, and piñon pine, which eventually fade into tender alpine fauna: moss and lichens above the tree line. As with many places in the West, overgrazing has forced the plant life to almost entirely repopulate, forcing old species away and allowing new species to therefore emerge dominant.

Continuing south and past the Henry Mountains, UT 276 passes through **Ticaboo.** Ticaboo is an unicorporated town with little more than a post office—so don't expect any kind of resupply here, fuel or otherwise. This highway then approaches the northern edge **Lake Powell** at **Bullfrog.** Those who wish to see the lake may head a few extra miles south along 276 to the water's edge and the visitor center. (And those wishing to bypass this stop may just turn west onto the Burr Trail, just past Ticaboo). Bullfrog itself is a touristy, lake-oriented town whose primary economy is servicing those who require a boat rental or a marina for putting their own boat into the water. As boating here is a heavily seasonal activity, don't expect much at all to be open from late fall through early spring.

Those suddenly feeling the urge to be on the south side of Lake Powell can take the **Bullfrog Ferry** to Halls Crossing and to a continuation of UT

AMERICAN BISON

A heard of roughly 500 of these buggers live in the Henry Mountains. This special group is one of only four genetically pure, free-roaming (i.e., non-captive) herds in North America. The other three are found in Wind Cave National Park, South Dakota; Elk Island, Alberta; and the famous Yellowstone National Park, Wyoming and Montana.

Prior to Europeans arriving permanently in North America, these animals traveled in massive herds across the continent, from near-central Mexico to the upper northwestern section of Canada. Populations at that time have been estimated to be in excess of 50 million; today, with the aid of preserves, the count is somewhat less than 500,000—with the majority living as "owned" animals.

This herd hasn't simply survived to this date in the Henrys; it was intentionally seeded there in 1941 by the transplant of 18 bison from Yellowstone Park to the southeast of Hanksville. As the herd grew naturally, so too did it shift its home to the Henry Mountains.

Biologists believe that the herd consists of roughly 175 animals too many—so today animals are being transplanted to other areas in the region where, introduced to other bison, they hopefully will create new herds. Additionally, lottery-style, once-in-a-lifetime hunting permits are granted yearly. Because of the extreme topography and ruggedness of the terrain, this is not an easy shot.

276. The ferry is only in regular service April 1 through October 31, save for winter maintenance runs. While operational spring through fall, it runs four to five times a day, with the schedule changing easily. No reservations are possible, but all planning to use the ferry should call ahead to check the schedule and arrive early to assure themselves a spot. Warning: To enter the **Glen Canyon National Recreation Area,** a $15 toll must be paid (national parks passes accepted).

As the Burr Trail heads west toward Boulder, it first winds and climbs gradually through natural drainages, nearing the **Waterpocket Fold.** As the road gets closer to this feature, it goes through this maze of buttes and washes, gradually climbing toward the centerpiece of the drive: 3 miles of road with switchbacks of incredible severity. As the road approaches the Waterpocket Fold, it will become increasingly obvious why the Burr Trail—which navigates through this barrier—is a significant and famous route. The likelihood of passing a standard vehicle through any point of this feature seems silly without massive blasting and modern construction, both of which were unavailable to John Burr. And though it is not recommended that RVs or vehicles with trailers pass through here, any normal passenger vehicle should be fine. Just shift into low gear so as not to burn out the transmission. Over the course of the drive, the road climbs in elevation from 3,900 feet to 6,600 feet above sea level.

After climbing to the top of the last switchback, the road becomes much tamer. Drivers will be able to relax, shift out of low gear, and take in the view. Look south (left) to see the Circle Cliffs in the distance. With an

---

JOHN ATLANTIC BURR: NAMESAKE OF THE BURR TRAIL

From Bullfrog Marina to Boulder, Utah, the **Burr Trail** is a 66-mile scenic byway that passes though the **Waterpocket Fold, Capitol Reef National Park** (no fee to cross through; there are no park facilities here), and **Long Canyon.** This route is named after pioneer John Atlantic Burr, a homesteader who built a 3-mile stretch of this road, the most unlikely part of today's current throughway. Born in 1846 on the Atlantic Ocean (hence the middle name), aboard the S.S. *Brooklyn,* John moved west and eventually made a life for himself driving cattle between Bullfrog Basin and Boulder, Utah. Crossing through the Waterpocket Fold—a 100-mile-long, formidable rock wall—required the construction of a path through a weakness. Requiring a passageway through which to drive his cattle, this is what John Burr created, earning a legacy for his name.

The Muley Twist Section of the Burr Trail

elevation up to 7,251 feet, they rise about 600 feet above the plateau at road level. From the right vantage angle (looking back toward the east), the Henry Mountains form a deep blue backdrop behind this striking red cliff band.

The eastern front of the Waterpocket Fold marks the eastern border of **Capitol Reef National Park.** For the next stretch, this drive will traverse through an "unimproved" section of this park. Though signs alert passers-by to the park boundaries, no fee must be paid. Instead, drivers can pull off along the way and picnic at one of the roadside tables on the plateaus atop the plateau ascended by the Burr Trail.

The last major feature—and minor topographical obstacle for the car—is **Long Canyon,** which the road dips down into, then arrives out of, just east of Boulder. Very soon before the descent into Long Canyon, there is a good view of the Circle Cliffs Amphitheater with the Henry Range as a backdrop. At this point, the road to Boulder is just less than 18 miles long.

The next bit of pavement meanders its way along the flat bottom of a very narrow, ehm…long canyon—about 7 miles in length. With shorter walls and a much more slender girth, this somewhat resembles the Colorado River east of Moab, with its sheer, smoothly sculpted Navajo sandstone walls—burnt red with black streaks. To the north is Rattlesnake Bench, and to the south is King Bench.

About 5.5 miles beyond the top of the Long Canyon, look for a small side canyon to the right (north). This abrupt, non-technical slot canyon gives observers a taste of such environs without the effort or technical gear required to access most in the area. Dark and usually a bit damp, this corridor is much cooler than the environs, so don't let ambient heat deter you from an opportunity to explore off-road.

The mouth of Long Canyon is just more than 10 miles from the outskirts of Boulder. This cute, artsy desert town and Escalante are prime stops in the Grand Staircase-Escalante National Monument—replete with coffee shops, B&Bs, bookshops, galleries, and restaurants. Bustle on up to Boulder, take out the guidebook, and find a good place to cozy up for the night. (For more information on lodging, dining, recreation, and shopping in this area, check out chapters 11 and 12 of this book.)

WATERPOCKET FOLD

Yet another of Utah's kooky and unique geological freak shows, this is the defining feature of Capitol Reef National Park. A hundred miles long, this sandstone reef is actually a massive fold in the earth's crust, also called a monocline, or "step up" in rock strata. About 60 million years ago, massive tectonic activity in the area released movement in an older fault, allowing the rocks around it to shift. This movement resulted in the west side of the fault being lifted 7,000 feet higher than the east side, and the earth on that side folding over the fault itself.

All of this occurred underground, but as the Colorado Plateau was lifted higher and higher in the last 20 million years, the ground layers atop this fold were finally washed and blown away. Subsequent weathering has revealed many colorful layers of stone. The name "waterpocket" refers to dips and basins in the rock that erode as a result of water erosion—a similar idea as the water tanks in Hueco Tanks State Park, Texas.

IN THE AREA

Attractions and Recreation

Bullfrog Visitor Center, Lake Powell and UT 276, Bullfrog. Open irregularly, beginning in May. Call to inquire at 435-684-7420.

Hanksville-Burpee Quarry, Hanksville. Call District Manager, Todd Christiansen, 435-586-2401. Web site: www.blm.gov/ut and www.burpee.org.

Glen Canyon National Recreation Area, P.O. Box 1507, Page, AZ. Call 928-608-6200. Web site: www.nps.gov/glca.

Goblin Valley State Park. Visitor center open 8–5 daily (sometimes closed for staff lunch breaks in the off-season). $7 day-use fee; $16 per campsite. Campground reservations: 1-800-322-3770. Web site: www.stateparks.utah.gov/parks/goblin-valley.

The John Wesley Powell River History Museum, 1765 East Main Street, Green River. Admission is $4 for adults 13 and up, $1 for children. Open daily 8–7 in summer, Tuesday–Saturday 9–5 in winter. Call 435-564-3427. Web site: www.johnwesleypowell.com.

Dining

Blondie's Eatery & Gift Shop, 40 North UT 95, Hanksville. Diner-style restaurant where customers order at the counter. Open Sunday–Thursday 7–8, Friday–Saturday 7–9. Call 435-542-3255.

Blue Mountain Pizza Deli, 30 East UT 24, Hanksville. Take-out only; seasonal hours. Call 435-542-3249.

Ray's Tavern, 25 South Broadway, Green River. Call 435-564-3511.

Other Contacts

Bullfrog Ferry, End of UT 276, Bullfrog. In regular operation April 1–October 31; no reservations. Call 435-684-3088.

Escalante River

CHAPTER

11

Escalante to Capitol Reef

Estimated length: 72 miles from Escalante to Capitol Reef National Park. From there, another 37 miles to Hanksville (to join the Henry Mountains route, as detailed in chapter 10); or, alternately, another 73 miles to Richfield via UT 24, which brings the ride to a close at I-70. Extra detours suggested as well.

Estimated time: Roughly 3 hours from Escalante to Hanksville, or 3.5 hours to get from Escalante to Richfield. With any sightseeing, this drive should take a day; many detours could occupy weeks.

Getting There: This drive begins in Escalante—the endpoint for chapter 12's route, "Parowan to Escalante." That drive (along UT 12) is the fastest and best way to approach from the south; if coming from the north, arrive by way of UT 24, which connects to I-70 in Sigurd and west of Green River.

Highlights: Along the way are two of Utah's coolest desert towns, **Escalante** and **Boulder**—gateways to the area's hiking, canyoneering, and mountain biking. Vibrant, yet remote, these cater to visitors, but have not yet been choked with tourism. Along the way are multiple museums and state parks to illustrate the history of the **Grand Staircase-Escalante National Monument.** The final destination on the route is the famous **Capitol Reef National Park.**

Any drive that begins in **Escalante** is off to a good start. This little oasis of fun and artsy-ness barely occupies a dot on the map, yet for a town with a population less than 900, it has a rich concentration of inviting cafés, restaurants, and guest beds. Though modern-day Escalante is a fun stop along a great highway, it hasn't always been so accessible; UT 12 was built only in the 1950s, replacing the notoriously hairy Devil's Backbone road.

Even in Utah, this particular region stands out as a less-traveled corner throughout history. In 1776, Franciscan explorer-priests Escalante and Dominguez engaged in a famous northward trek from present day Santa Fe, New Mexico. They hoped to determine an overland route to Monterrey, California. Though this mission failed under approaching winter in northern Utah, when the priests were forced to return home, they nevertheless accomplished the first major exploration of the continental interior.

Other Europeans subsequently entered the region in passing, making note of it. During the Black Hawk War of 1866, Captain James Andrus—for some reason—named it Potato Valley. The famous Green River explorer, Major John Wesley Powell and his head mapmaker, A.H. Thompson, visited the area in the 1870s, documenting its topography and ecology.

On one trip in 1875, Thompson encountered a group of Latter-day Saints attempting to establish a township there. As the story goes, he advised them to name the town Escalante after the early priests. And so it was done. Built by Mormons, this town indeed features the gridded and number-named streets that follow this religion's trademark "Zion Plan" for city layout. Original brick homes from this period still stand between noticeably wide streets.

Early area residents enjoyed mild winters and survivably hot summers. They made a living in cattle and sheep ranching, as well as mining and logging. Long growing seasons and endless sunshine allowed for irrigated crops to be plentiful. This town served as a crucial Mormon outpost those coming from Salt Lake City en route to colonize the area in the name of the Later-day Saints church. Even given the presence of towns such as Escalante, these groups still had their work cut out for them when it came to overland travel.

The town of Escalante has grown and shrunk—but its current population of 818 is just a few hundred less than its 1940s all-time high. Isolated from major cities and roadways, this town enjoys its prime location on the incredibly scenic **UT 12**. This highway streams a tourist population through Escalante during the summer, on the way to and from **Bryce**

HOLE IN THE ROCK ROAD

In the fall of 1879, a group of colonizing Mormons set out from Salt Lake City en route to what would become Moab. Called the San Juan Mission, this expedition sought a route across Utah to its southeastern corner. Moving to an unsettled region, they carried as many supplies as necessary in cumbersome wagons. Even today, with modern roads and vehicles, Utah's terrain is steep and difficult to pass. However, back in the 1870s, the region was almost entirely uncharted, and the group certainly had no access to modern roads or vehicles.

In the name of taking the most direct route possible, the group traveled across the relatively flat **Kaiparowits Plateau** beneath the **Straight Cliffs.** This worked out well until they encountered what were then 1,200-foot-tall cliffs at the northern edge of the **Glen Canyon.** Needless to say, this was a pretty "stopper" feature. Not to be turned away, scouts began exploring the canyon rim to find a weakness. Searching revealed the "Hole in the Rock," an extremely narrow and steep slot cutting down through cliffs and toward the **Colorado River** (now **Lake Powell,** penned behind the **Glen Canyon Dam,** with a much higher water level). Immediately across the river, they spotted the Cottonwood Canyon, leading back up and onto the **Wilson Plateau.** Convinced, they prepared to cross the river via these two features.

However, the Hole in the Rock needed alteration before it could provide adequate passageway for wagons, livestock, and people. Explosives widened the necessary parts of the slot, and wagon-lowering-anchors were carved into the stone. The group implemented wooden tracks and sandstone slabs to smooth sections of the slot, enough that wagons and livestock could successfully arrive at the river, down grades as steep as 50 percent. In early 1880 they made it to the river, only to find the way up Cottonwood Canyon to be even more challenging.

Once atop Wilson Mesa, the group of 250 people, 1,000 animals, and 83 wagons traveled onward and established the town of Bluff in 1880. This Hole in the Rock route was used for just a year by incoming residents and suppliers, before a much more reasonable route, passing through Halls Crossing, was discovered and put into use.

Canyon and **Capitol Reef** national parks, the **Burr Trail** (in chapter 10 of this book), and many other state parks.

For grub in Escalante, look no farther than UT 12, which swoops through town as **Main Street.** About half a dozen restaurants stand there, trying to lure in business with Western-themed names and décor. Cow-

Navajo Sandstone by the Escalante River Trail   Caitlin Hutter

boy Blues, for example, serves ribs, steak, trout, pasta, pizza, burgers, soup, salad, wine, and draft beer—daily and all year. Casual and family friendly, they offer a special children's menu.

Georgie's Corner Café and Deli, probably the most popular among locals and passersby, tends to keep its lights on longer than other places and serves—in addition to burgers, sandwiches, and desserts—a reputedly good Mexican selection. Georgie, who relocated from California, apparently brought that state's famous Mexican tradition with her and offers very full plates to hungry guests. Operating in a converted house, this restaurant has a cozy, homey atmosphere. Especially during the busy season, arrive for dinner early; located in a small, isolated town with few options for resupply, this popular restaurant does run dry on ingredients.

For coffee, pastries, sunscreen, fishing and hiking gear, guidebooks, a place to camp, or even a bed to sleep in, check out the **Escalante Outfitter /Esca-Latte.** This cute little shop/café is like Escalante's clown car of services. Though the front building has a small appearance, it's charmingly packed with a huge variety of retail and rental goods to match the vast selection of recreation in the Grand Staircase-Escalante National Monument. Additionally, there's a restaurant that serves pizza, quiche, and more. Tucked behind and around the shop is the lodging and camping portion of the campus. The owners and staff here tend to be quite cheery and do everything, from cleaning the cabins, to making espresso drinks! They're a hard-working crew—even offering guided fly fishing trips and informational hikes, as well as tips and recommendations for bike rides.

Cabins, bunks, and campsites are available for rent at Escalante Outfitters, but if they're full, several alternatives exist right in town—and, just as with the dining options, these are located primarily on Main Street. Large groups or families (up to 12 people), should consider staying at the **Red Rock Ranch.** This single 2,700-square-foot "cabin" sits at the southern end of town. With generous windows and a hilltop position, this two-story affair provides plenty of space for groups not wanting to split up. Originally a family home, this has all the amenities for a self-sufficient vacation, like a kitchen, fireplace, and lawn. Built in 2000, this has a clean, modern interior.

For a couple, check out **Canyons Bed & Breakfast.** Lacking in the batty, old grandmother feel that a lot of bed and breakfasts exude, this has no overstuffed beds, lace curtains, or Victorian anything. Instead, this 1905 ranch-style house gives guests private room entry (with covered porch) and modern, yet simple, newly remodeled rooms. Breakfasts consist of well balanced, home-cooked dishes, and include fresh fruit, eggs, and coffee. (For those wishing to get an early start on the day, continental breakfast is available before the official meal.) A limited number of rooms—just four—keeps the establishment uncrowded.

Those with a few spare hours to check out something beyond the hotel room should visit the **Escalante Petrified Forest State Park.** First, as the name would suggest, a sizeable petrified forest is the showpiece of this park. Two foot paths meander through the grounds, allowing visitors to get a closer look at fossilized tree remains, some of which are greater than four feet wide. One, the Petrified Forest Trail, is roughly a mile long and climbs 200 feet over its course; the other, Sleeping Rainbow Trail, is a quarter mile

Sunrise on the Hogback UT 12 between Escalante and Boulder Caitlin Hutter

shorter and is an extension of the first. A third option, Baily's Wash, can be walked—but is not a developed trail; consult a ranger to find its location and get directions about the actual route.

In addition to the hiking trails and petrified forest there is…a lake? Yep, the **Wide Hollow Reservoir.** Created in the 1950s to provide irrigation for Escalante, it is presently stocked with rainbow trout and bluegill. Though the reservoir was drained for a dam reconstruction project in March 2010, it is expected to be open after that year. To get there, head slightly west from downtown on UT 12/Main Street and then north on Reservoir Road. The onsite campground provides an alternative to in-town lodging.

Now comes the part where you must choose your own adventure: either take UT 12 toward Boulder, checking out its own attractions and detours on the way (**Hole in the Rock** and **Calf Creek Falls**), or head north out of Escalante toward **Posey Lake** and the famous **Hell's Backbone Road.** We'll start with the UT 12 version, but then come back and visit the Posey Lake/Hell's Backbone alternative.

Escalante to Boulder via UT 12

Though the famous Hell's Backbone route certainly takes the breath away, this trip to Boulder on UT 12 certainly isn't boring either—providing ridgeline driving and plenty of scenery along the way. A section of this route called **The Hogback,** about two-thirds of the way to Boulder, is a neatly curving section of UT 12 atop a ridge with steep drop-offs on either side that certainly encourage the driver to remain attentive. About 29 miles in length, it should take just less than 40 minutes complete without detours.

Those interested in checking out some wild pioneer history should visit **Hole in the Rock** (see sidebar). This narrow notch was the passageway that allowed late 19th century settlers to navigate the then 1,200-foot cliffs down to the shores of the Colorado River. (Today, the river is dammed and the water levels of Lake Powell are much higher than that of the unaltered river.) Evidence of this early civil works project remains clearly visible

Calf Creek Caitlin Hutter

Peekaboo Canyon, Hole in the Rock Road Caitlin Hutter

to this date: blasting marks left by explosive powder, anchor points for rope-lowering wagons, and even steps to smooth the way.

To visit this remarkable couloir, head east from Escalante (about 4.5 miles) on UT 12, turning right (southeast) on **Hole in the Rock Road.** This dirt lane leads to the national historical site after about 55 miles, passing through an explorer's paradise on the way. Though well maintained for about the first third, the road surface grows less and less smooth, eventually becoming impassible by non-off-road vehicles somewhere near the end (between 5 and 15 miles, depending on grading). Therefore, the last section of road makes for a great, low-skill mountain bike ride. Along the Hole in the Rock Road, many side roads branch out, offering hiking trails, wild camping spots, and picnic areas. If heading into this area, it's worth getting a somewhat detailed map, so as not to miss these lesser-known opportunities.

Back on the main road to Boulder, consider stopping at **Calf Creek Falls,** a popular attraction in the area. Accessible by hiking only, this 126-foot waterfall sits atop a trail whose starting point is the Calf Creek Recreation Area Campground (15 miles east of Escalante on UT 12). Campsites here cost a small amount and come with pit toilets, picnic tables, and fire rings. The namesake attraction, Lower Calf Creek Falls, is reached by hiking just more than 2.5 miles up a relatively flat, yet sandy and narrow canyon. **Calf Creek** runs year-round with clear, cool water, falling into a scenic pool in a cozy sandstone amphitheater. Even though the pool is cool and refreshing, the hike is, like the rest of the area, hot in summer and should be done with sufficient fluids and sun protection. (The Upper Calf Creek Falls, upstream from here, can also be accessed, but from a separate slickrock trail that departs from UT 12 5.5 miles north of the campground/main trailhead.)

Escalante to Boulder via Posey Lake and Hell's Backbone

Posey Lake Road can be picked up easily in Escalante; it starts as 300 East Street (a north-south running lane) in the eastern side of town. Take this street north, following it as it doglegs east and then north again and out of town and turns into Posey Lake Road. Stay on this route for a total of about 18 miles and take a right on Hell's Backbone Road, for roughly 20.5 more miles of driving. This dumps directly onto UT 12 again, about 4.5 miles south of Boulder. The total route to Boulder is about 44 slow miles on dirt road, suitable for any passenger vehicle (barring severe weather).

With the route plan sorted, get ready for one of Utah's more famous sections of driving. Posey Lake Road climbs north from Escalante and up into the high Dixie National Forest. **Posey Lake** itself, about 15 miles north of town, is enclosed by a mixed aspen and evergreen forest. At 8,600 feet above sea level, its campground provides a cool, shaded rest stop. Hiking trails departing from the Posey Lake lead to Tule Lake and connect with segments of the Great Western Trail; numerous dirt roads crisscrossing the forest pass by other lakes, and can be casually mountain biked. Though the elevation is quite high here, the topography is nevertheless pretty mellow. There is small overnight fee for this campground, which features a boat ramp (for non-motorized boats only) and pit toilets.

The **Hell's Backbone Road** itself was built by the Civilian Conservation Corps in 1935 and reaches an elevation of up to 9,000 feet above sea level. This high, exposed road thrills visitors as it traipses along crumbling, forested ridges and spiny sandstone mountains. Snaking around, it gains popping vistas down and across the region. Along the way is the famous **Hell's Backbone Bridge,** which penalizes inattentive drivers with a 1,500-foot plummet if they err. Though a gravel road, it is a popular tourist attraction and therefore well maintained. People often describe the road as terrifying to drive—but truly a proper road, it requires no special technical skills to navigate.

Arriving in Boulder—a nexus in the Grand Staircase-Escalante National Monument—requires visitors to regroup and decide where to go next. From here, many directions of travel are possible. But don't be too slow with decision-making, as the drive through town can nearly be missed in the blink of an eye. To take the **Burr Trail** (detailed in chapter 10 of this book), head east out of town on the **Burr Trail Road.** Easily located, this road departs east from town exactly where UT 12 makes a sharp, 90 degree town to the north (left) upon its entrance into Boulder. Taking this road commences a reverse-direction version of chapter 10's tour of the Burr Trail and the **Henry Mountains.**

However, remaining on this chapter's prescribed route, one of the first and most easily accessible attractions in town is the **Anasazi State Park Museum.** Here, a village dating from roughly 1050 to 1200 A.D. has been partially unearthed. The intactness of the structures allows for easy imagining of the lives of these Ancestral Puebloan people who managed to forge a settlement here 1,000 years ago. From the roughly 100 structures has been recovered a variety of artifacts suggesting the type of lifestyle these folks had. These items can be seen on display in the museum, which also serves as a **information center and permit vendor** for the Grand Staircase-Escalante National Monument.

Another immediately obvious attraction in Boulder is the **Hell's Backbone Grill,** the region's best restaurant. Focusing on fresh cuisine, it serves home-grown vegetables whenever possible—which is usually the case, given the immense growing season that just overlaps nicely with tourist season. When not grown on site, the produce served is nonetheless organic, and the meats come from local sources. The area's most acclaimed restaurant, the grill is located on expansive, gardened grounds that pleasantly embellish the dining experience. Hell's Backbone Grill has claimed

GRAND STAIRCASE-ESCALANTE NATIONAL MONUMENT

This enormous monument occupies 1.9 million acres (nearly 3,000 square miles) of land. About 1.5 times the size of Delaware, it sits in the south-central portion of Utah. To the south, it is bordered approximately by US 89 and Lake Powell; on the east, it ends at Capitol Reef National Park, and on the west it stops roughly at Bryce Canyon National Park and the Dixie National Forest.

Spanning about 9,000 vertical feet and roughly two-dozen major rock strata, this lifted and folded piece of crust has revealed some of the earth's most beautifully deposited rock layers as it has cracked and been eroded by wind and water. Forming the "staircase" are the Pink, Grey, White, Chocolate, and Vermillion cliffs, as they step down from an altitude of 12,000 feet near Cedar Breaks and the Paunsaugunt Plateau, down to 3,000 feet above sea level at the bottom of the Grand Canyon. Fissured and sculpted into its steps are Bryce Canyon, Zion and Kolob Canyons, the Sevier River, and the Coral Pink Sand Dunes.

This vast and wild landscape offers an incredible wealth of recreation and scenery, in addition to natural and human history. Unlike other national monuments, this is administered by the Bureau of Land Management instead of the National Park Service. Designated a national monument only in September of 1996 by Bill Clinton (during his presidential reelection campaign), it is has been speculated that this monument was created perhaps in an effort to gain votes. Some officials complained about the sudden allotment of such a vast portion of land, concerned with lost mining opportunities and access; others were glad as, even though a national monument does not enjoy identical protection wilderness lands, its sanctity is much more sheltered than it would be without national monument status.

To learn more about Grand Staircase-Escalante National Monument, stop by any of its offices or visitors centers, of which there are more than 10. See the Web site for more information. And be careful, as always, to check the weather before exploring slot canyons to be sure that there are no storms systems in the area; flash floods kill people every year in Utah.

multiple Utah State best-of titles, as well as national awards and write-ups. Located in southern Utah, the atmosphere here is casual.

For an equally soul-fulfilling slumber, make reservations to stay right here at the **Boulder Mountain Lodge.** Situated on the same green property as the Hell's Backbone Grill, this hotel maintains its space from the restaurant, but enjoys the same environmentally conscious ownership and

Boulder Mountain Quaking Aspen Caitlin Hutter

wholesome, relaxing surroundings. Spacious rooms with private balconies overlook the grounds and environment beyond. Pets are actually welcome here. Due to the magnetic nature of both the restaurant and the lodge, would-be guests should be sure to lock down reservations well ahead.

For a fewer-frills, fewer-dollars dinner, check out the **Burr Trail Grill.** This serves a selection of house entrees, sandwiches, burgers, salads, and soups, as well as beers and wine. Next to the grill and deli (in the same building) is the **Burr Trail Outpost,** which brews espresso drinks, sells guidebooks and gear, showcases local artwork, and arranges guide services. This restaurant is open seasonally for lunch and dinner only; call ahead if visiting during the shoulder season.

And for a more affordable, just-want-to-sleep-in-a-bed experience, check out the **Circle Cliffs Motel.** Located right on Main Street (UT 12), toward the northern end of town, Circle Cliffs is definitely a no-frills motel, and nothing fancy. Rooms range from single bed to apartment. A place for those who'd rather invest their money in other parts of the trip.

Heading north from Boulder, visitors say goodbye to the Grand Staircase-Escalante National Monument and look forward to a 35-mile trip to **Torrey** and UT 24. Heading more or less due north, UT 12 enters the **Boulder Mountain** section of the drive. Not a single dot on the map, this "Land of 1,000 Lakes" offers itself up to hunters, fishers, hikers, campers, and backcountry explorers. This massive mountain, which tops out at 11,317 feet above sea level, is a wild area for diffuse recreation. Numerous forest roads reach west into the region, accessing lakes, campgrounds, and general wilderness. High aspen stands and deep green, conifer forests contrast with red, sandstone valleys of the surrounding area. Lakes here are often crowned with encircling cliff walls and forests, thinned somewhat by their high elevation. The best method for exploring Boulder Mountain begins with choosing the category of recreation. Then grab a detailed topographic map, stock up with supplies, and head into the wild.

Because of all the lakes and streams, fishing seems to be one of the most popular activities in the area. **Boulder Mountain Fly Fishing** happily escorts those without their own equipment, knowhow, or wherewithal in the region. These guys know the ins and outs of the area's streams and lakes and how they respond to season, time of day, and weather. In a region with so many bodies of water—literally thousands—including teeny, often unmapped lakes, beaver ponds, and small streams, this kind of expertise can be quite helpful. All necessary gear will be provided (if required). All persons fishing in Utah are required to have a license; check on Boulder Mountain Fly Fishing's Web site for a link to Utah's license vendors.

Time to boogie out of Boulder Mountain, up past the small town of Torrey, and onto UT 24, heading east to eventually reach Capitol Reef National Park. But those who missed out on dining in Boulder, should try Torrey's **Café Diablo.** In a casual atmosphere, this restaurant serves a creative hybrid menu, including contemporary American cuisine, Southwestern, and South American dishes, with fish, poultry, steak, and even rattlesnake. Not just a meat-and-potatoes kind of place, the choices here excel in creativity, with homemade sauces and fresh, unique sides. Despite a laid-back dining room, the menu is high-end.

Just east of Torrey is **Capitol Reef National Park,** called a "reef" as its major feature—a 100-mile-long spine of rock—forms a veritable barrier for any east-west land travel. The **Waterpocket Fold,** a Navajo sandstone ridge, is a huge uplift that when viewed aerially, appears to be a stiffly pleated portion of the earth's crust, with parallel ridges of pale yellow and deep red sandstone. From the ground view, it can easily be understood why early pioneers found it to be a legendary barricade to travel.

The Waterpocket Fold formed sometime around 60 million years ago when local tectonic activity produced a monocline; a monocline or "step" in the earth's crust, is a fold so severe that strata, which would otherwise sit horizontally, have been tipped on edge. At one point, the lifted side of this fold was more than 7,000 feet above the lower. However subsequent fracturing and erosion has disrupted the original surface, buffing it down and exposing the many colors of the vertically tipped strata. The "Waterpocket" portion of its name comes from the nature in which this feature erodes. As water eats into the steep sandstone via narrow slits in the rock, it creates pool-like "pockets" in the sandstone.

Two primary sections of the national park exist: Capitol Reef and Cathedral Valley. The **Capitol Reef** portion gets its name from the golden Navajo sandstone domes that resemble government buildings—and reef for the obstacle they present to land travel. About 10,000 feet and 250 million years' worth of rock layers can be seen here. First lifted by the Waterpocket monocline, and then eroded by subsequent time, these strata can be seen anywhere along this fold.

As any national park does, Capitol Reef includes a paved, scenic drive that covers the visual highlights of the park without requiring an exit from the car. This primary road departs from UT 24 and heads 10 miles south

McGath Reservoir on Boulder Mountain Caitlin Hutter

into the park. The first sight along the way is **Fruita,** a bygone orchard and a historic home, the **Gifford Homestead,** which dates back to the early 20th century. Here period artifacts are on display and lifestyle demonstrations take place.

Unlike many parks, however, most of Capitol Reef's road miles remain unpaved; beyond the scenic drive asphalt stretch many other long, looping dirt roads that should be driven with and accompanying gazetteer or park map. About 4 miles south from UT 24, the dirt **Grand Wash Road** heads east from the pavement, twisting for a mile into that wash. Where the road ends, a trail begins, and leads deeper into more scenic stretches of Grand Wash. Along the walk, the trail passes **Cassidy Arch,** named after outlaw Butch Cassidy of famous Wild West history, who occasionally sought refuge from the law here. This sturdy, round arch forms a window from a plateau above, down to the narrow gulch below. Also along the gulch are small, manmade caves; these small tunnels in the rock are the remnants of defunct Uranium mines from the early 1900s, and are a very common sight around the desert southwest. A word of caution: stay away from this road and hike altogether if there is any bad weather in the area; as a wash, this serves as a natural drainage way, and flash flooding can easily arise without warning.

At the end of the scenic drive, the pavement ends, and the 10-mile road must be retraced. It is advisable to stop in at the visitor center and pick up a park guide before embarking on this drive; as the road tours so many millions of years' worth of geological history, many interesting things can be seen in the rock that would go unseen without a few tips. With some simple education, the eye can pick out rock that was deposited by wind and water—and fossilized footprints can be picked out as well.

The northern region of Capitol Reef National Park, or **Cathedral Valley,** must accessed by an entirely different road system to the north of the park's UT 12-accessible section. Located in the far northern corner of the Capitol Reef, it can either be reached by dirt road heading northwest from Cainville—or by heading east from UT 72, also via dirt road, into the park. The Cathedral Valley is oft unvisited, but contains some of the most beautiful features to be found within the park. Inside a broad, flat, and green valley, stands a city of red rock fins and spires quite resembling cathedrals. Before heading for a visit to this valley, call the park to determine the current road conditions; these can become worn-out and very rocky, requiring high clearance. Also, wet conditions transform the dirt into a very slippery and sticky clay cable of miring vehicles.

After Capitol Reef National Park, the way out is up to the driver: continuing east along UT 24 bring travelers directly to **Hanksville,** which sits near the beginning of the Henry Mountains/Burr Trail route (chapter 10 of this book); heading west on UT 24 leads toward I-70 near Richfield, and back onto fast roads that quickly access civilization.

IN THE AREA

Accommodations

Boulder Mountain Lodge, 20 North UT 12, Boulder. Moderately priced (cheaper during the off- and shoulder seasons). Pets welcome. Open year-round. Call 435-335-7460 or 1-800-556-3446.

Canyons Bed & Breakfast, 120 East Main Street, Escalante. Affordable. Open March through October only; closed in winter. No children under 12 permitted; pets not accepted. Call 435-826-4747 or 1-866-526-9667. Web site: www.canyonsbnb.com.

Circle Cliffs Motel, 225 North UT 12, Boulder. Very affordable. Open year-round. Call 435-335-7333.

Red Rock Ranch, 450 East 300 South, Escalante. Moderate to pricey; prorated fees for longer stays. Single-home rental for large groups; five bedrooms. Call 801-619-6434. Web site: www.redrockranchescalante.com.

Attractions and Recreation

Anasazi State Park Museum, 460 North UT 12, Boulder. Small fee for entry. Open April 1 through October daily; November 1 through March 31 Monday–Saturday; some holiday closures. Call 435-335-7308. Web site: www.stateparks.utah.gov/parks/anasazi.

Boulder Mountain Fly Fishing. Call 435-335-7306 to arrange trip, meeting place. Web site: www.bouldermountainflyfishing.com.

Capitol Reef National Park, 11 miles east of Torrey on UT 24. Mailing address: HC 70 Box 15, Torrey 84775. Call 435-425-3791. Web site: www .nps.gov/care.

Escalante Petrified Forest State Park, 710 North Reservoir Road, Escalante. day-use fee. Open year-round; some holiday closures. Camping available (for an extra fee). Call 435-826-4466. Web site: www.stateparks .utah.gov/parks/escalante.

Grand Staircase-Escalante National Monument, Kanab Headquarters, 190 East Center, Kanab. Call 435-644-4300. Web site: www.ut.blm .gov/monument.

Dining

Burr Trail Grill and Outpost, 10 North UT 12, Boulder. Casual atmosphere, American and Southwestern cuisine. Call 435-335-7503. Web site: www.burrtrailgrill.com.

Café Diablo, 599 West Main Street, Torrey. Moderately priced. Indoor and outdoor seating; casual. Reservations recommended; open mid-April through mid-October nightly for dinner only. Call 435-425-3070. Web site: www.cafediablo.net.

Cowboy Blues Restaurant, 530 West Main Street, Escalante. Affordable. Western American cuisine. Open daily for lunch and dinner, year-round. Call 435-826-4577. Web site: www.cowboyblues.net.

Escalante Outfitters/Esca-Latte Café, 310 West Main Street, Escalante. Gear retail, bike rentals, pizza restaurant, beer, and espresso. Guided informational hikes and fly fishing trips. Cabins, bunks, and campsites for rent. Call 435-826-4266 or 1-866-455-0041. Web site: www.escalante outfitters.com.

Georgie's Corner Café and Deli, 190 West Main, Escalante. Hours change seasonally; call ahead. Call 435-826-4784.

Hell's Backbone Grill, 20 North UT 12, Boulder. Moderately priced. Contemporary American cuisine crafted with organic and local produce and meats. Reservations strongly recommended. Open mid-March through Thanksgiving. Call 435-335-7464. Web site: www.hellsbackbone grill.com.

Other Contacts

Dixie National Forest, presiding over the bulk of the land along this route. Call 435-865-3700. Web site: www.fs.fed.us/dxnf.

Escalante Outfitters (see Escalante Outfitters/Esca-Latte Café, above)

Interagency Visitor Center in Escalante, 755 West Main, Escalante. Information on the many nearby state and national parks, as well as Grand Staircase-Escalante National Monument. Call 435-826-5499.

Bryce Canyon Arch Caitlin Hutter

CHAPTER

12

Parowan to Escalante

Estimated length: 115 miles from Parowan to Escalante (plus 40 extra miles round-trip to include the Parowan Gap Petroglyphs and Grosvenor Arch detours)

Estimated time: 2.5 hours driving; days or weeks with stops and exploration

Getting There: This area is about as packed with as much geological beauty as possible—from tall, evergreen-covered mountains, to exposed red rock canyons. That said, there is no specific "best" way to get here. The main goal is to get into the vicinity of Parowan (exit 75 off I-15), departing from the interstate north of Cedar City on UT 143. Those with extra time and arriving from the north, should consider driving US 89 instead of I-15; this runs roughly parallel to the interstate but tours Circleville Canyon alongside the Sevier River.

Highlights: Look at the atlas and notice that this entire route is dotted with "Scenic Drive" markings—and there's no wonder why. Here the earth has been more uplifted, gnarled, colored, and scraped bare than almost any other place in the country. Over just 50 miles in the first section of this route, UT 143 ascends 4,500 feet (nearly one vertical mile) and through half a dozen major life zones. This pavement ribbon connects some of the state's most scenic areas as it enters the western rim of the glorious and sizeable **Grand Staircase-Escalante National Monument.** Along the way, it passes **Brian Head Ski Resort** and summer playground, **Cedar Breaks National Monument, Bryce Canyon National Park,** the turnoff for **Kodachrome Basin State Park,** and **Escalante State Park** in the town of Escalante. Though this drive "ends" there, the next (detailed in chapter 11 of this book) begins from the same point (and even connects into chapter 10's drive to the Henry Mountains).

Parowan, Utah, seems but a dot on the map, with a population of just over 2,600. Surprisingly, this was one of the first towns incorporated in Utah, and the first in all of southern Utah. Founded in 1851—just four years after the Mormons' arrival in Salt Lake City—it hasn't grown much.

The original inhabitants of the area are thought to be the Fremont and Anasazi Indians, and petroglyphs dating from 750 to 1250 A.D. serve as evidence. To check out these very striking and detailed drawings, make a short detour to the **Parowan Gap Petroglyphs,** just northwest of town by way of Parowan's **Gap Road.** Onto the faces of varnished sandstone boulders (on the gap's east side) have been etched many complex images. These drawings signify aspects of prehistoric life, from hunting to spiritual rituals.

Because of the illustrations' variety of age, style, and content, many think these markings were the work of several different cultures, likely

---

PAROWAN: WORLD RECORD LAWN-MOWIN'

Most people in Utah don't know this, but this small town is home to a *Guinness Book of World Records* record-holder for...the longest lawnmower ride. In 1997 resident Ryan Tripp saddled his riding lawnmower in Salt Lake City and set out east, driving 3,116 miles east to Washington, D.C., raising $15,000 for the kidney transplant of a young girl. The kicker? Ryan was only 12 years old at the time of this journey. The lawnmower itself was not modified in any way; it was quite a wide, very normal sit-down lawnmower. Afterward, he appeared on *The David Letterman Show* and made a few other media appearances.

Two years later, he journeyed to each of the 50 state capitals in order to advocate for tissue and organ donation awareness. This time, he did not travel on his mower; but he did bring it with him, mowing the lawn at each state's governor's mansion.

---

including the Columbian Fremont, Sevier-Fremont, the Southern Paiute, and more. In any case, dating shows these etchings to span roughly a thousand years, and depicted figures include snakes, lizards, mountain sheep, humans, and bear paws. The most famous and studied item, though, "The Zipper" is believed by many to represent a time/space calendar that records the motion of the sun over time.

To get there, head north on Parowan's Main Street to 400 North Street. Turn left (west) and continue for 10.5 miles on a high-quality gravel road, parking near Milepost 19. **Nal Morris** is one of the primary researchers of the pictograph's cosmological significance; visit his Web site (www.parowan gap.org) for more information.

Moving onto the route proper, leave Parowan (the town), heading south on UT 143. Roughly 8.5 miles south of town, look for a pullout (and wooden fence), as this denotes the trailhead for **Hidden Haven Falls.** This small interpretive trail heads into an ever-tightening, highly pleasant, vegetated canyon. Nearly all of the 1.5 miles of trail are user-friendly and hikeable by anyone; only the last 100 feet are rugged and difficult to pass. The first part of the trail heads through a wildlife study area and is consistently signed. Sharing the same parking area is the namesake falls, where a waterfall cascades roughly 200 feet. By winter, this freezes and is frequently visited by ice climbers.

Brian Head ski resort and summer playground, a little more than 4 miles beyond the falls. During the snowy months, this small resort has

PAROWAN GAP

Looking to the west, one must wonder how there came to be such a distinctive notch in the Red Hills. Before the existence of this ridgeline, a stream ran through what was then a valley. Tectonic activity in the area, however, caused a long strip of sedimentary rock to cut free from the surrounding crust. Pressed between two parallel faults, it was lifted upward by squeezing. As this band of rock rose, the stream continued its course, cutting a gap through the rising earth faster than it could grow. Eventually the climate changed and the stream disappeared. Today wind continues to carve away at this weakness, rushing by as it funnels through this narrow gap. If you go to check out the petroglyphs, take notice of the elevated wind speeds in the 600-foot-tall notch.

the best skiing in southern Utah. By summer it offers lift rides, disc golf, a hiking trail (new in 2010), and mountain biking. Mountain bikers can come with or without their own wheels; a shop right at the resort rents out a variety of bikes, including cross-country and downhill frames, as well as children's bikes and accessories. Like the bicycle fleet, the mountain's selection of trails consists of cross-country and downhill options, intermediate to advanced trails, short rides and long, technical and (relatively) leisurely. Shuttles and lift services for bikers can be purchased. Helmets are required, and can also be rented, along with "armor." Disc golf equipment rentals are available as well.

About a mile south of Brian Head on UT 143 sits a forest service road heading east. Signed by the Dixie National Forest as BRIAN HEAD PEAK 3, this road leads to a lookout point. A high-quality, 2.5-mile dirt road climbs to the top of **Brian Head Peak,** elevation 11,307 feet. Despite the lofty elevation of this peak, its shape is quite rounded and gentle. Atop the mountain stands a small wood-and-stone hut, built by the Civilian Conservation Corps between 1935 and 1937. From the top of this mountain, several ranges can be spied. To the south stand Arizona's Navajo Mountain and Mount Trumbull, Nevada's Highland Peak and Wheeler Mountain, and Utah's Tushar Mountains, Table Cliffs, and the Paunsagunt and Aquarius plateaus; to the west are Utah's Wah Wah and Never Summer mountain ranges.

Hardly a few miles farther south on UT 143 (at its junction with UT 148) comes the next stop, **Cedar Breaks National Monument.** One of Utah's most remarkable natural excavations, this orange basin cuts deeply and abruptly into a high, green plateau. This bowl-shaped ring of brightly colored, striated rock formed by steep erosion, leaving behind 2,000-foot curved walls of crumbly, fire-red earth. With its bold colors and peculiar weathering patterns, it undeniably bears a strong resemblance to Bryce Canyon National Park—but may even feature richer and more vibrant colors than its sister.

As long as 900 years ago, the Southern Paiute lived in this area. This tribe was based in southern Utah, and lived somewhat nomadically. Planting valley crops in the spring, they would then climb high into the mountains in and around present day Cedar Breaks to get some reprieve from the severe summer heat of lower elevations.

Within the park is a handful of picnic areas, campsites, and two hiking trails for use by summer visitors. The road, however, typically becomes impassable for the winter anywhere between October and December, depending on the year. Yet rather than shutting the park, the closure actually opens the road up for cross-country skiers, snowmobilers, and snowshoers. This is one of the most scenic spots for Nordic skiing in Utah, as the only thing that contrasts the red walls more beautifully than green forests is winter's white snow.

A Winter Warming Yurt stays open during the cold months and substitutes as an impromptu visitor center. The normal visitor center is open, and the fee booth active, only from roughly June 5 through the middle of October; but the scenic drive remains open as long as it stays free of deep snow. As when visiting any high elevation locale near winter, be sure to check ahead to be sure that all roads are open; its 10,000-foot altitude make Cedar Breaks a catcher and holder of deep snow pack.

Moving east along UT 143, keep an eye out for small, high mountain lakes. The next section of geology, vastly different from that of Cedar Breaks, features lava flows—just a few thousand years old and clearly visible on the way to Panguitch Lake. At this point, the road rides along the highest step on the so-called **Grand Staircase** (see sidebar, below), a massive geological feature centered around the Four Corners region of the United States. Look around from this perch and, on clear days, the view extends for hundreds of miles with Arizona visible to the south beyond the descending Grand Staircase.

Once at **Panguitch Lake,** ahh—take a breath of thick air, for the eleva-

GRAND STAIRCASE?

The so-called "Grand Staircase" is a geological set of steps set inside the huge— 130,000 square miles—Colorado Plateau. This region of the Southwest is centered around the Four Corners intersection of Arizona, New Mexico, Utah, and Colorado. Serious tectonic activity along the Hurricane Fault (which trends east-west south of Cedar City) and the Sevier Fault (another east-west fault north of Kanab) yielded this spectacular uplift with more than 6,000 vertical feet of relief. As this rapid uplift occurred, hundreds of millions of years of sedimentary deposits were revealed to expose sandstone that would oxidize to become deep and vibrant reds, oranges, and yellows.

The lowest step of the staircase lays in the south near Kanab, Utah, and Lake Powell, and ascends in major rock strata. These Chocolate, Vermilion, White, Grey, and Pink cliffs. Here, Bryce Canyon has eroded into the topmost of these, the Paunsaugunt Plateau; Zion National Park, to the southwest, has been created by the Virgin River eroding steeply into a Navajo sandstone layer sitting atop the White Cliffs.

tion is but a paltry 8,400 feet on the Markagunt Plateau! With very little tall vegetation surrounding this broad **Dixie National Forest** lake, it has quite an open feel. Named by the Southern Paiutes, "Panguitch" rougly means "big fish," and indeed sizeable rainbow trout can today be fished from these waters during the summer months. During the cold winters here, fishermen erect ice fishing huts.

The strong presence of rainbow trout actually came about as the entire lake was treated with rotenone, a broad spectrum pesticide and insecticide, to remove a non-native chub species. As this agent kills all fish, and not just the targeted, non-native chub, the lake had to be restocked with 20,000 rainbow trout in 2006. A natural lake by origin, this had an original surface area of roughly 780 acres, but that was expanded by about 60 percent with the installation of a dam.

Infrastructurally, the Panguitch Lake environs are pretty much undeveloped. Several campgrounds open during the summer for those who want to stick around for a while and explore. If hungry, snoop around, as convenience stores and a few lodge-style restaurants keep visitors and locals fed and happy. Some of these sit on the West, North, and East "Loop Roads"—a paved detour that, together with UT 143 encircles the lake. Because of the open, unpopulated nature of the Markagunt Plateau, this corner of Utah lends itself to snowmobiling, cross country skiing, and snowshoeing on backroads during the white months.

Less than 20 miles northeast of Panguitch Lake, the highway swoops to the northeast, and then ends in US 89. At this junction sits

Harris Wash Caitlin Hutter

the actual town of **Panguitch,** elevation 6,660 feet. Joining the National Register of Historic Places in November 2006, this settlement can indeed claim some beautiful and well preserved brick homes as well as some truly cool, old western-style storefronts. Panguitch offers more amenities in the way of lodging and dining than any other in the Bryce Canyon region.

Main Street is littered with motel-style lodging. Locally owned, these modest accommodations thrive on the business brought to them by their location in "Canyon Country"—but many open only May through October. A few B&Bs can be found in town as well. One of these, the **Red Brick Inn,** a 1920s-era Dutch Colonial home, offers a more homey stay with a family. Guests enjoy evenings sitting on the porch, or making s'mores at the fire pit in the back yard. Located right in heart of this quiet town, it requires no detour to reach.

BRYCE AMPHITHEATER NATIONAL PARK

Technically not a canyon, Bryce is actually an amphitheater. Whereas canyons, by definition, have been carved top-down by streams eating into parent rock, Bryce was eroded from the bottom up, as low waters ate away at the plateau above. Regardless of geological definition, this particular area of Utah might stand alone as one of the most unique and visually striking geological features in the state.

The dining, like the lodging in Panguitch, exudes small mountain town flavor. Most eateries can be found along Main Street, are easily spotted, and serve lots of meat. A couple of café-style restaurants offer coffee and causal meals; a few national-brand chains also offer quick bites. Look around and take your pick. Before leaving town, be sure to fuel up the car as townships from here on out are small and far between.

Next up is **Bryce Canyon National Park,** south of Panguitch on US 89 for 7 miles, and then southeast on UT 12 toward Boulder. UT 12 offers much more than just a means to get to Bryce; it is one of the state's most stunning roadways, touring a major section of the **Grand Staircase-Escalante National Monument.** Here it enters this national monument atop the tall and expansive Paunsaugunt Plateau, blanketed in high-altitude evergreen and sage. This highland flows uninterruptedly until suddenly—as if it were the edge of the earth—it falls steeply away, exposing thousands of vertical feet of strangely eroded, brilliantly colored sandstone layers. Glowing orange, white, red, and yellow, this special area came under protection as a national monument in 1923, and later a national park in 1928.

Bristlecone Pine on Powell Pt near Bryce Canyon Caitlin Hutter

In comparison to many of Utah's other spectacular sandstone formations, Bryce's features are quite different. Its walls are not narrow and tall like those in Zion National Park; it has no buffed Navajo sandstone windows as in Arches, and it lacks the lofty Wingate towers and fins of Canyonlands National Park. Rather, this broad, scooped-out bowl is filled with thousands and thousands of soft, spiny-looking sandstone spires that resemble red sugar drippings.

Among this peculiar forest of rock stand peculiar features that are weirder yet—the **hoodoos.** These buggers are recognizable as thin rock pillars topped precariously with broader capstones and boulders. Resembling humans or even aliens, these form when sturdy layers of rock sit atop more erosive layers. The tougher capstones wear away less quickly than their supporting strata. Despite the unlikely nature of these hoodoos, they can sometimes reach heights of 200 feet. Though spires and hoodoos are the main theme of this park, a few stone arches and fins do exist here, however stylized with Bryce Canyon's knobby flavor.

Amazingly, given the local desert climate, different peoples have lived in the area for roughly 10,000 years. Artifacts from the Basketmaker and Pueblo-period Anasazi have been recovered in the area. The Paiute entered the region roughly 800 years ago, leaving behind artifacts of their own, including petroglyphs tapped into varnished sandstone faces.

The first Europeans descendants to live in the immediate vicinity were the Mormon family of Ebenezer Bryce, who ran a small cattle ranch at the base of the "canyon" during the 1870s. Persistently inhospitable conditions—including harsh winters, spring flooding, summer drought, and self-caused overgrazing—forced the family to throw in the towel after just a few years, and relocate to Arizona in 1880. For more information on geology, history, or ecology of the region, visit the **Bryce Canyon Natural History Association,** which runs a bookstore inside the park's visitor center. Created in 1961, this non-profit assists with educational and scientific activities within the park.

Leaving Bryce Canyon, continue along UT 12, now trending east/southeast; next up is the turnoff to **Kodachrome Basin State Park,** named so by the National Geographic Society for the fact that Kodak films replicate the wonderful colors of this area: reds and yellows. (Its classical competitor, Fujifilm, typically excels with greens and blues).

Just when you think Utah couldn't possibly offer any more variety in its strange sandstone earthscapes, along comes Kodachrome. Here, 16 peculiar cylindrical sandstone spires, or "sedimentary pipes," jut high into the sky from stone platforms atop flat, grassy fields. It should be immediately obvious to the observer why this has been designated a state park—by all appearances, it is nature's own sculpture garden with bright colors, spires, arches, and other peculiar formations backdropped by a basin of brightly colored earth. In the roughly 5,000 acres of the park, visitors can camp (for a fee), hike, mountain bike, horseback ride, or simply relax. To reach Kodachrome Basin state park, look for **Cottonwood Canyon Road** departing to the south (right) from UT 12, about four miles south of Tropic; follow signs to the park from here.

Grosvenor Arch, 10 miles east/southeast of the Kodachrome Basin (farther along Cottonwood Canyon Road), should be considered a worthy detour if time isn't limited. This distinctive, gothic-looking double arch has been carved naturally by desert winds. From its top to bottom, the red rock fades into pale tan as it nears the earth. The pullout for the arch should be easy to spot, as a short sidewalk leads directly to its base from a pit toiled and picnic tables. The National Geographic Society, just as with Kodachrome Basin, named this arch after its one-time president, Gilbert Hovey Grosvenor.

Toodle back north to UT 12, and continue in a northeasterly direction as it crosses into **Grand Staircase-Escalante National Monument** just before arriving in the town of **Escalante.** GSNM and Escalante are both showpieces of the region; with the national monument encompassing 1.9 million acres of land, and Escalante serving as an outpost for desert enthusiasts to decompress and restock.

Though this drive "ends" there, the next route (detailed in chapter 11 of this book) also begins right here. From Escalante, it continues through the Grand Staircase-Escalante National Monument, past Boulder and Capitol Reef National Park, and eventually up toward I-70.

IN THE AREA

Accommodations

Red Brick Inn, 161 North 100 West, Panguitch. Open May–October; reservations accepted year-round. Very affordable. Call 435-676-2141 or 1-866-733-2745 Web site: www.redbrickinnutah.com.

Attractions and Recreation

Brian Head Resort, 329 South UT 143, Brian Head. Open in winter and summer. Call 1-866-930-1010. Web site: www.brianhead.com.

Bryce Canyon National Park, P.O. Box 640201, Bryce Canyon 84764. Day-use fee per vehicle (with an exception on certain government-determined days, see Web site); national parks pass accepted; additional fees for camping. Call 435-834-5322. Web site: www.nps.gob/brca.

Cedar Breaks National Monument, intersection of UT 148 and 143, Brian Head. Entrance fee per individual visitor over the age of 16; national parks pass accepted. Call 435-586-9451. Web site: www.nps.gov/cebr.

Kodachrome Basin State Park, (mailing address:) P.O. Box 180069, Cannonville 84718. Day-use fee per vehicle; additional charges for camping. Call 435-679-8562. Web site: www.stateparks.utah.gov/parks /kodachrome.

Other Contacts

Bryce Canyon Natural History Association, P.O. Box 640051, Bryce 84764. BCNHA is a non-profit organization established in 1961 to assist with the scientific, educational, and interpretive activities within the park. Call 435-834-4782. Web site: www.brycecanyon.org.

Dixie National Forest. Presiding over the bulk of the land along this route. Call 435-865-3700. Web site: www.fs.fed.us/dxnf.

Grand Staircase-Escalante National Monument, Kanab Headquarters: 190 East Center Street, Kanab. Call 435-644-4300. Web site: www.ut .blm.gov/monument.

Nal Morris, expert on Parowan Gap petroglyphs. Web site: www.parowan gap.org.

Panguitch Lake Visitor Guide. Web site: www.panguitchlake.com.

Weeping Rock

CHAPTER

13

St. George to Zion National Park

Estimated length: 50 miles each way
Estimated time: A few hours with no hiking

Getting There: St. George, Utah sits at the far southwestern corner of the state, right on I-15, and just north of Arizona's Virgin River Gorge. St. George is just about 120 miles of interstate northeast of Las Vegas, and a very quick 300 I-15 miles south of Salt Lake City. Given speed limits between 75 and 80 miles per hour, the trip from Salt Lake should take less than four hours.

Those coming up from the south on I-15 (and through the Virgin River Gorge) should be advised that this is an exceptional piece of freeway. Much windier that most, this very narrow canyon climbs quite steeply and twistily, with reduced speed limits for a reason. There is no cell phone service in the gorge, so be sure to tank up on gas in Mesquite, Nevada, before attempting to ascend this stretch of road.

Highlights: St. George is Utah's retirement community for a reason. Surrounded by beautiful, red rock desert, it enjoys a nearness to Las Vegas, both in terms of distance and climate. Yet it has its own, much more conservative culture and small town accoutrements. In and around St. George are endless opportunities for recreation, from golf and biking, to hiking and climbing in **Snow Canyon State Park.**

Reaching **Zion National Park** takes less than an hour and, though it requires some driving on I-15, most of the drive takes place on relatively scenic, low-traffic highways. **Springdale,** just before Zion's west entrance, serves visitors with lodging, gear rentals, and restaurants. And Zion National Park is yet another of Utah's spectacular and unique geological phenomena. Carved by the **Virgin River,** this deep, narrow canyon has sheer sandstone walls sometimes in excess of 1,000 feet, towering above a pleasant, sandy, cottonwood-tree-lined canyon bottom.

St. George is the southwestern-most city in the state, barely within Utah's boundaries. Its nearest major attractions are **Zion National Park, Cedar Breaks National Monument, Brian Head Resort** and, a bit farther afield, **Bryce Canyon National Park.** Despite its nearness to these destinations, St. George is hardly a tourist town; it has a thriving local economy of its own, based largely in retirement, recreation, and construction. Headquartered here are Sky West Airlines, Dixie State College, and a major hospital. Located precisely on I-15, which runs directly from Salt Lake City to Las Vegas, it is easily accessible. Yet, roughly 120 miles from Las Vegas and about 300 from Salt Lake City, it has an identity and character all its own.

Since the 1848 Mormon inception in Salt Lake City, this southern region of Utah has always had the name "Utah's Dixie" or just Dixie. Within a few years of establishing roots in Salt Lake City, the Mormons had already begun colonizing more widely in the region. By 1851, the first Dixie settlement in present-day Parowan, called the Iron Mission, had been established by a group of 168, led by Parley Pratt. By 1857, Harmony, Santa Clara, and Washington (just north of, and touching St. George) had all been settled. Dixie's roots were lain. Not a fly-by-the-seat-of-the-pants group, the Mormons calculated each new development, beginning each city with an organized infrastructure, including an irrigation plan and street layout. Since that time, each of these communities has fleshed out, and more have filled in around them.

In recent decades, the population St. George and Washington County has been growing remarkably quickly. Between 1990 and 2000, it actually grew almost a percent more than Las Vegas, edging it out for fastest-growing metropolis in the United States for that period, and earning the second place title in 2005 (behind Greeley, Colorado). From 2000 to 2009, the population grew more than 52 percent to roughly 140,000 residents.

Census Bureau experts have projected a population more than 700,000 by 2050 if nothing changes between now and then.

For a tasty bite of fresh ingredients, generous proportions, with yet plenty of flavor and salt to quench post-exercise cravings, what could be more fulfilling that a Thai/sushi/Asian fusion combination restaurant? A bit pricier than peer restaurants in larger cities, **Benja Thai and Sushi** is located right in historic downtown, in Ancestor Square, among boutique shops and brick walkways.

If leaving town in the daytime and wishing for a quick lunch, coffee, or breakfast, café-style, stop in at **25 Main.** A fleshed out, creative, and espresso-serving café this dishes up breakfast, panini, salad, pizza, and famous cupcakes. Each week has a set cupcake schedule, with different varieties being offered every week, and multiple choices each day; check online for the schedule.

Mountain bikers should know that the roads between St. George and Zion are virtually crisscrossed by some the state's best, most scenic bike trails—much more user-friendly than their slickrock counterparts near Moab. Additionally, the side roads themselves can make for a great on-pavement bike trip. Those planning to ride, and needing to rent a bike, might as well do it in St. George. **Red Rock Bicycle** has a rental fleet with men's and women's options in both road and full-suspension mountain bike fleets. In addition to the rentals, there is a retail side with a full selec-

---

SNOW CANYON STATE PARK

Those who haven't seen the famous Red Rocks Scenic Loop (west of Las Vegas), and don't necessarily have to time to get there, should check out **Snow Canyon.** Though not technically a part of this chapter's drive, Snow Canyon is virtually in St. George and is therefore hardly out of the way.

Unlike many state parks around the nation, this is extremely well maintained and sparkly clean. Cozy in feel, it nevertheless offers some of St. George's best multi-pitch rock climbing, picnicking, hiking, biking, and camping. Roughly 18 miles of trail exist within the park so, though it's not an endless backpacking destination, it provides well more than a day's worth of exploration. The scenery is dominated by orange Kayenta and Navajo sandstones topped with black basalt. Just about 7,500 acres, this park is actually a subsection of a more than 62,000-acre Red Cliffs Desert Reserve.

---

tion of accessories from tubes to pumps, sports foods, maps, and locks—even trunk and hitch racks for hire. Even those with all necessary bikes and accessories should consider a stop here to visit with the staff and get tips on the best, most suitable trail or road ride for the circumstances. The most famous rides are **Gooseberry Mesa, Hurricane Rim,** and **J.E.M.; Snow Canyon State Park** has good riding, too. The first three (and the most noteworthy) are more or less right on the drive into Zion.

To leave St. George, simply get to I-15 and head north. Not long after entering the interstate does the route depart via exit 16 to the well signed UT 9. This route busts dead-east to **Hurricane** at first, and then begins squiggling up through **La Verkin** and **Virgin** before coming to **Rockville** and eventually **Springdale.** This final town abuts directly to the western entrance of the park, just about 30 miles past I-15. Along the way, this fairly serious highway climbs slopes and dips into valleys as it approaches Zion.

Along this drive eastward, the area's many mountain bike trails can be picked up by way of side roads. Finding and following these trails is fairly intricate, so it's best done with a map and more detailed directions. One of the most popular means to find these trails (and others all across the state) is the free Web site, www.utahmountainbiking.com. This long-standing, amazing resource offers mini-guides and route descriptions per region, supplemented with photos, videos, GPS coordinates, route lengths, and elevation gain. Based on personal experience, the descriptions can sometimes have errors; but keeping alert and following intuition can avoid derailment.

SAND HOLLOW STATE PARK

Yet another short detour, this little state park sits just south of UT 9. Historically it's been popular among ATVers, dirtbikers, and dune-buggyers who ride its sand dunes. A new reservoir (as of 2003) now attracts fishermen and seals the deal on its designation as a state park. Beaches and picnic areas make this family-friendly. Utah's newest state park, this was dedicated in 2009. Open for day use and camping.

After heading through **Hurricane** (pronounced "Hurricun" by locals), UT 9 cuts across the desert, ascending quite distinctly in places as it gains higher benches. Approaching Rockville and Springdale, it winds along the Virgin River bottom, with cottonwood trees growing out of the red sands, and crumbling buttes on either side. **Rockville,** the last town before Springdale has remained a quiet, residential town with little intention to forge a business in tourism. Its

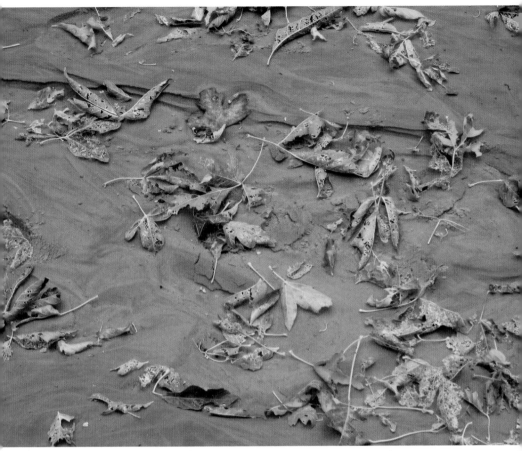

Leaves in Virgin River Sand

old-fashioned, genuine charm almost causes second-guesses as to whether Zion is actually ahead. Even the street lights here are antiquated, with single bulbs hanging from wires suspended across the street.

Up next is **Springdale,** the obvious gateway to Zion with restaurants, hotels, outdoors shops, and cafés galore—all dressed in desert, outdoorsy motifs. A few places stand out as worthy pit stops. A quick pick-me-up, **Café Soleil** specializes in breakfast and lunch, but also serves lighter dinners. Equal parts restaurant and espresso bar, this café offers a generous selection in fresh, speedy eats. Breakfast can be light or hardy, with options in pastries, frittatas, breakfast sandwiches, and burritos. Lunchtime comes with wraps, panini, gyros, pizza, and smoothies. Dine in or outside, or take a boxed lunch to go.

For a hearty, yet gourmet dinner (or breakfast or lunch), try the **Switchback Grille**. Serving American and Continental cuisine, this particular establishment stands out for its fresh seafood, USDA Prime beef, wild game, wood-fired specialty pizzas, and free-range poultry. They even offer a cigar menu. Also a family-friendly place, there's a children's menu.

Heading east out of Springdale leads immediately to the west gate of **Zion National Park.** Carved by the **Virgin River, Zion Canyon** has been visited by people as far back as 8,000 years ago. The ancestral Puebloan, the Anasazi, came to be the first human residents in the area roughly around 6000 B.C. Small family groups of these seminomadic, basket-making people lived in the valley and traveled with the seasons to find food and livable climates. A subsequent group, the Parowan Fremont, lived there as well, but left around 1300 A.D., to be replaced by a completely separate group, the Southern Paiute. They arrived completely independently of their predecessors, and were the last residents of the canyon before the Mormon arrival to the area. In 1858, Mormon Euro-Americans first saw the area, led by a Southern Paiute guide.

A few years would pass after Zion's "discovery" before Mormons would actually settle here. In 1865, Isaac Behunin began farming corn, cotton, and tobacco in Zion Canyon proper. He and his family spent summers in the canyon (near the modern Zion Lodge), and would winter in a broader section of the canyon, where Springdale is today. Already by 1909, the land was recognized as special and designated as the **Mukuntuweap National Monument** by President William Taft. "Mukuntuweap" was its first European given name, when Major John Wesley Powell called it that on his 1872 visit, thinking it was its Paiute name. In 1916, though the name was changed to Zion, as its original name was unpopular among local

TANNER AMPHITHEATER

Fire up the computer and take a look at this venue's online schedule. During the summer, this **Dixie State College** facility hosts many musicians in a fabulous, open-air setting underneath the cliffs of **Zion Canyon.** Seating options include bench-style seats, as well as chairs with backrests. Each show usually features two or more acts, and the stage lights up almost weekly during the summer. Be sure to check the weather, though; there is no covered seating, so any rain falling from the sky will land on the audience. There is an indoor backup hall, but the experience there is much less special than the open-air treat of the main stage.

Mormons. On November 19, 1919, it was officially established as **Zion National Park.**

In its time, Zion has become the most popular park of all in Utah—more visited than each Bryce Canyon, Arches, Capitol Reef, and Canyonlands national parks and Monument Valley Navajo Tribal Park—receiving roughly 2.6 million visitors a year. To put it in perspective, Yosemite National Park, one of the nation's most famous and celebrated, gets just 3.5 million visitors a year—less than a million more than Zion!

Due to the enclosed nature of Zion and its limited roadways, the Park Service initiated a free shuttle system in 2000. These buses run between April and October every year, and are the mandatory means for transport within the park during that time. Arriving frequently and stopping at many places throughout the park, they actually seem to be an asset rather than a hassle, as they eliminate traffic jams and allow visitors to rubberneck without having to focus on the road. In the off-season, buses do run, though on a lessened schedule. If visiting during peak season, keep the eyes peeled, starting in Springdale for shuttle bus stops and associated parking areas. As most visitors are not local, the stops have been deliberately made readily visible and hard to miss.

One of the first points of interest inside the park is hardly past the fee station, reached by following the first right-hand spur. This is the **visitor center.** Interesting in its own way, this architectural gem has been cleverly designed to have the least environmental impact possible. Almost purely operational on just natural light and solar power, its heating and cooling tasks are accomplished with deliberately angled windows, cooling towers, and wall aspect and placement. Those interested can read up on it on plaques surrounding and adorning the building. Also at this site is the **Backcountry Desk,** which doles out permits to backpackers, rock climbers, and generally anyone who wishes to spend the night out and away from the main, pay campsites.

Right on this first turnoff is the **Watchman Campground,** one of only two in-park, Zion Canyon camping areas. The other is **South Campground,** accessed just a bit farther along the main drive (also to the east of the road). The South Campground driveway also serves as the access to the **Zion Nature Center.** Largely a curriculum-based organization, this center serves to educate mostly elementary school groups on the goings on of Zion Canyon and nature as a whole.

The next stop back on the main road (and to the left), is the **Zion**

ANGEL'S LANDING

Looking around the canyon, it's clear to see that these sandstone walls are absolutely sheer. Closer inspection also reveals cracks and other weaknesses running the entire height of many of these faces. Therefore, these cliffs are actually popular for rock climbers, both aid climbers and free climbers alike. What's the difference? Aid climbers actually pull gear to ascend the rock; free climbers use protection, but actually ascend purely by their own power, only using gear to protect a hypothetical fall. Both groups use stoppers and cams, but aid climbers also use hooks and any other device that can hold weight but likely not protect in the case of a fall.

Author with Climbing Gear

That said, the vast majority of people don't have the interest or technical skills required to climb these walls. Fret not, though. Reasonably fit persons can gain virtually the same head-spinning exposure simply by hiking the non-technical **Angel's Landing Trail.** This feature, located right in the "pinch" of the river's biggest bend, is a peninsula of vertical sandstone walls. To ascend Angel's Landing, this built-up trail makes many switchbacks, and is augmented by steps, chain hand rails, and the like. So, even though it's technically just a walking trail, it provides phenomenal excitement.

Human History Museum, opened in 2002. For those interested, this offers some visual displays and explanations of the various human groups that have lived in and been impacted by Zion. A free, short film narrates this story. Permanent and traveling displays of artifacts, photographs, writings, and replicas, illustrate the roughly 8,000 years of human habitation—from prehistory to Mormon "discovery" and its modern existence as a popular national park. At no cost, this should be considered a fully worthy stop.

The park's road system is quite simple, but soon it forks into two separate branches. One option heads to the north; the other goes east. Both are more fully described, below. The first, a continuation of the **Zion Canyon Scenic Drive,** heads up through Angel's Landing area, and eventually to the famous Zion Narrows. The second, the **Zion-Mt. Carmel Highway** passes through the famous **Zion Tunnel** that, part of a massive public works project, allowed this road to connect to the eastern end of the park. Finally open in 1930, the tunnel was then the longest of its kind in the United States. Outside of the tunnel, this highway makes many switchbacks along the way past the **Canyon Overlook Trail** and toward the eastern entrance for the park.

Zion-Mt. Carmel Highway

Heading east along the **Zion-Mt. Carmel Highway** leads quickly to the famous Zion Tunnel (in which no bike traffic is allowed). In the 1920s, the Park Service felt the increasing need to connect Springdale to the eastern end of the park; at that time, to reach Mt. Carmel (the town) or any of the surrounding villages, drivers had to circumvent the park in a large-scale way. Either to the south or to the north, it took hours to reach the town of Mt. Carmel. This passage, 1.1 miles in length, went into construction in the mid-1920s, and was finally completed in 1930. The opening up of this eastern passageway also renders Zion National Park a much more logical stop on the tour of southern Utah's national parks like Bryce Canyon and Canyonlands.

Along this road is the **Canyon Overlook Trail.** This short trail, just half a mile each way, climbs a good bit to reach generous views of **Zion and Pine Creek canyons.** Topping out just above the Great Arch, hikers on this trail cannot actually see this feature from the overlook. Beyond that, and just before the **East Entrance,** the **East Rim Trail** departs to the north, lacing around the **White Cliffs** and heading toward the northern end of the park, splitting into various trails after **Stave Spring** (for these options, it is best to refer to a detailed topographical map or a Utah gazetteer).

Zion Canyon Scenic Drive

To continue north and head toward Angel's Landing and toward The Narrows, drivers must turn left. About a mile north of the junction is the **Court of the Patriarchs** pullout, to the right. These "patriarchs," visible across the road (to the west), form an alcove of monstrous monoliths. They are also religiously named: Abraham, Isaac, and Jacob.

The next stop, on the right-hand side of the road is the **Zion Lodge.** This historic structure offers the only in-park, indoor lodging. Open all year, this gives guests the choice of hotel rooms, suites, and cabins. Also at the lodge is a gift shop and restaurant. Reservations are highly recommended for peak-season visits.

Speaking of this restaurant, it goes by the name of **The Red Rock Grill.** This eatery takes advantage of regional flavor and flare to embellish a bit on American cuisine. Fresh trout and sea fish, steaks, poultry, pasta, and the like, done well. The restaurant takes steps to improve its environmental responsibility, such as marking items made with sustainable ingredients, not automatically serving water, and selecting local or farm co-op meats and certified sustainable fish when possible. Like the lodge, the restaurant remains open year-round, and serves breakfast, lunch, and dinner. Dinner reservations are required!

Across the river from the Zion Lodge, trails depart to the **Emerald Pools.** Naturally set up for any level of fitness and agility, these three, tiered pools can be reached by hikes of increasing distance and elevation gain, from the lower, to the middle, and finally upper. The higher the trail climbs, the more rugged and steep it becomes—though it is still easily passable by able bodies. Though the stream that flows between the pools is not always running, the pools themselves always contain water, and the mossy, jungle-like environment they create in the desert is quite cool.

Though Angel's Landing hasn't been visible from the roadside yet, the famous trail departing to this formation begins at the **Grotto,** just half a mile up the road from the Zion Lodge. To access this hike, cross the Virgin River at the footbridge there and head north on the **West Rim Trail.** Prepare for some remarkable exposure and thrills on the way up! No simple foot path, this is rather a highly developed and reinforced construction with guard chains, cut steps, hand rails, and switchbacks. This particular trail is to be avoided in bad weather, particularly in high winds or thunderstorms. But any rain also renders the path slippery, which can be dangerous with these kinds of drop-offs.

As the road continues to the north, be on the lookout for rock climbers ascending the canyon walls—particularly where the Virgin River makes its S-curves in the Angel's Landing vicinity. This has some of the park's highest concentration of climbs, from single-pitch to dozen-plus-pitch routes. One of the more famous (and easily recognizable) climbs called **"Moonlight Buttress"** ascends a very sheer, smooth, pillar on the other side of the river, just northwest of the second bus stop in the S-curves. This was free-soloed (i.e., climbed with no rope or protection) in 2008 by a professional climber.

The final stop is at the very terminus of this road (and the driving route). Those who wish to extend the journey a bit may do so on foot by hiking into the **Zion Narrows**, where Zion Canyon pinches down to a size much too small to be shared by the road and river. Though a popular hike, it is certainly not a trail; this

Beneath Moonlight Buttress, 9 Pitches

route follows the actual Virgin River. That said, this is best hiked in late summer or fall when the temperatures are still warm, but the water is low. A sturdy, strap-on sandal is the best footwear as usually about half of the journey involves wading or even swimming. About 16 miles of river can be walked before the canyon becomes no longer passable. Just as Angel's Landing provides great exposure to non-climbers, this provides a semi-slot-canyon experience for non-canyoneers; along the way, walls can be as high as 2,000 feet, with the canyon floor as narrow as 30 feet. It is extremely important to avoid this hike if there is any precipitation in the area, even out of sight; storms upstream can cause flash flooding totally inescapable in such a narrow, sheer-walled canyon. Inquire with the park rangers at the visitor center if there is any doubt.

To return to the St. George area, simply head back exactly along this route. But to continue on and check out more drives or national parks, head east out of the park on the **Zion-Mt. Carmel Highway.** In **Mt. Carmel** (just outside of the park), UT 9 joins with US 89. This road heads north and from here, it is possible to join up with chapter 13's drive (Parowan to Escalante), and then farther on, chapter 12's drive (Escalante to Captiol Reef, and beyond).

IN THE AREA

Accommodations

Zion Lodge, about 2 miles north of the park's Springdale entrance, on Zion Canyon Scenic Drive. Pricey. Guest rooms, suites, and cabins available; reservations highly encouraged. Call 435-772-7700 or 1-888-297-2757. Web site: www.zionlodge.com.

Attractions and Recreation

Sand Hollow State Park, 4405 West 3600 South, Hurricane. Admission and camping fees. Day-use approximately restricted to sunlight hours. Call 435-680-0715. Web site: www.stateparks.utah.gov.

Snow Canyon State Park, 11 miles west of St. George by way of Bluff Street, 3.5 miles farther west on Snow Canyon Parkway, and then a signed right for Snow Canyon Drive. Open daily, year-round; no holiday closures. Daily admission fee. Call 435-628-2255. Web site: www.stateparks.utah.gov.

Tanner Amphitheater, west end of Lion Avenue, Springdale. Outdoor venue with no rain cover; mixed benches and seats with backrests. Call 435-652-7994. Web site: www.dixie.edu/tanner/index.html.

Zion Human History Museum, 0.5 miles into Zion National Park, on the west side of the Zion Canyon Scenic Drive. Call 435-772-3256.

Zion National Park, Springdale. Call 435-772-3256. Web site: www.nps .gov/zion.

Dining

Benja Thai and Sushi, Ancestor Square, 2 West St. George Boulevard, St. George. Cheap to moderately priced. Serving Thai and sushi. Open nightly for dinner. Call 435-628-9538. Web site: www.benjathai.com.

Café Soleil, 205 Zion Park Boulevard, Springdale. Very affordable. Open daily for breakfast, lunch, and dinner. Call 435-772-0505. Web site: www .cafesoleilzion.com.

The Red Rock Grill, in the Zion Lodge, about 2 miles north of the park's Springdale entrance, on Zion Canyon Scenic Drive. Moderately priced. American cuisine with regional ingredients and flare. Reservations required for dinner; open year-round, three meals a day. Call 435-772-7760. Web site: www.zionlodge.com.

Switchback Grille, 1149 South Zion Park Boulevard, Springdale. Appropriately pricey. American and Continental cuisine. Open for breakfast, lunch, and dinner daily. Kids menu offered. Dinner reservations encouraged. Call 435-772-3700. Web site: www.switchbacktrading.com.

Twentyfive Main, 25 North Main Street, St. George. Serving breakfast, espresso, panini, salad, pizza, and famous cupcakes. Very affordable. Call 435-628-7110. Web site: www.twentyfivemain.com.

Other Contacts

Red Rock Bicycle, 446 West 100 South, St. George. Open daily, with shortened hours on Sunday. Road and mountain bike rentals and retail. Call 435-674-3185. Web site: www.redrockbicycle.com.

Utahmountainbiking.com. Free mini-guides and route descriptions with photos, GPS coordinates, ride lengths, elevation gain, and more.

View Southeast from the Shoulder of Notch Peak

West Desert: Delta to Great Basin National Park

Estimated length: 105 miles (each way)
Estimated time: A short half day; a full day with return trip

Getting There: Delta sits just 30 miles west of I-15 as the crow flies. If coming from the north, take UT 32 south from Nephi, linking up with US 6. Southern arrivals should depart from I-15 in Holden, heading northwest on US 50.

Highlights: An utterly desolate and lonely stretch of highway, US 6 and 50 stretches from Utah's main north-south corridor, I-15, though the forgotten West Desert, and into Nevada. With views utterly uninterrupted by civilization, this landscape of dry lake beds and alternating rugged peaks has an otherworldly feel. Ending at **Great Basin National Park**—just on the other side of the Utah-Nevada border—this route provides a peak at the geologically unique **Great Basin** and a chance to imagine what Utah was like before it was irrigated and paved.

Unlike most townships in Utah, **Delta** wasn't settled until after 1900—and also unlike most townships, it came about as a result of agriculture being brought to the area, and not by colonizing efforts of the LDS Church. Alfalfa farming was already in full swing nearby to the east, fed by the **Sevier River** running through the valley between the **Wasatch Plateau** and the

Pahvant Range. A group of 34 local entrepreneurs became interested in the potential farming opportunities to the west and organized themselves in the effort of diverting water from the Sevier River westward.

A site was selected for the township in 1906, and settlers began laying roots. Three-hundred-and-twenty-acre tracts of land were doled out under the provision that they become irrigated within a three-year time period. A diversion dam was built, financed by stockholders, and interested families began moving in.

By the 1920s, Delta was one of the largest alfalfa producers in the nation, and had erected a sugar factory as well. Accompanying businesses were founded, and people continued arriving by way of the nearby railroad. The town weathered the Great Depression quite well, and its cattle ranching industry grew during this time. Mining of fluorspar was added to the list of commerce. Also called fluorite, this mineral has a variety of uses and, depending on the grade, can be used in smelting metals, reducing chromatic aberration in lenses, and in the production of hydrofluoric acid.

Given all this, present-day Delta is about as exciting as its name. For eats, there is a collection of fast food dispensaries, convenience stores, and one supermarket. The **Rico Taco Shop,** however, is Delta's coolest, local restaurant. Starting very early each day (5:30 AM!), they dish up classic American and Mexican breakfasts; in the evening, out come the tacos, mariscos, enchiladas, and combination platters—burgers and tortas, too. A kids menu is offered, as well as a salad list. And they broadcast free Wi-Fi all day.

Also in Delta, the **Great Basin Museum** is a small-town affair, but one that details some of the unobvious tidbits of information on the area. Worth a stop, as there is no charge. On display are fossils, geological specimens, and human history artifacts dating from prehistory to early settlers, all illustrated with local photographs and writing. Also featured in the museum is a darker chapter in the country's history, the **WWII Topaz Relocation Center**

NO SERVICES NEXT 100 MILES

Before leaving town, tank up the car; roughly 100 miles of civilization-barren pavement stretches from Delta to **Baker, Nevada,** just north of Great Basin National Park. It'd be a strange day when someone else didn't drive by, but, having actually had car problems on this road, first-hand experience says that it might be a while before a ride back into town avails itself. Also know that most things in rural Utah don't open on Sunday, including mechanic shops.

(also called Central Utah Relocation Center). Here 8,500 Japanese Americans were detained from 1942 through 1946, largely hailing from the San Francisco area. The museum holds various vestiges from this camp.

Heading out of Delta, drive west right on Main Street (also US 6 and 50). The road goes through the even smaller township of **Hinckley.** Take a quick detour to the south on UT 257 to **Fort Deseret,** located near mile marker 65 (on the southern end of Hinckley and about 5 miles south of Delta; located right off the highway). In 1865 during the Blackhawk Indian Wars, this fort was built by almost 100 men, for the purpose of protecting white settlers in the region. As the U.S. Government was quite involved in the Civil War, this was conceived of and erected by Mormons.

Opened after just 18 days of construction, this 550-foot square structure with 10-foot-high walls successfully protected the area's people and livestock when Chief Blackhawk arrived in the spring of 1866—a time of predicted hostility. After the threat passed, this structure's function shifted to be a cattle pen. What remains today is a fairly large adobe structure whose gun ports are still in place in the outer walls.

JOHN W. GUNNISON MASSACRE

In the fall of 1853, John Gunnison was appointed leader of an exploratory mission to the west-central part of Utah. His team's duty was to survey the region, making topographical maps and gathering data on plant and animal life and geology. Gunnison, an army engineer and West Point alumnus, knew Utah well and was quite fit to lead this team. Totaling 11 men, this group included a botanist, topopgrapher, artist, guide, and more.

This research journey, however, fell right in the heat of the of the **Walker Indian War,** named for the ongoing battles and skirmishes involving Chief Wakara of a band of Utes, the Pahvant Indians. In light of this, the Gunnison group had been warned that the natives might be hostile. Having traveled past **Fillmore,** they reached a point along the Sevier River, just southwest of present-day Delta, and made camp. At night, a troop of what appeared to be 30 Pahvant warriors descended on this group, killing all but four men.

However this story is not as simple as it sounds. Martha Gunnison, widow of Captain John Gunnison, believed that the attackers were not Utes of any kind. Rather, she maintained that the attack was carried out by a group of militant Mormons, the **Danites** (see sidebar, below). She believed this order was given by Brigham Young himself, who is alleged to have feared the advent of the **Transconinental Railroad,** lest it deliver more non-Mormon settlers to the area.

Martha wrote to the Utah's Federal Supreme Court appointee, Justice W. W. Drummond, who agreed with her based on witness and informant testimonies and depositions. Accounts by Pahvants and white settlers alike convinced Drummond that Mormons dressed as Indians had attacked this group. Brigham Young, Utah's chief for the Bureau of Indian Affairs, arrested the alleged Ute attackers, but never punished them. It is alleged that Brigham Young cut a deal with the Indians, paying them to take the blame for the crime. Of course, intervening time, deceased witnesses, and lack of records render this all hearsay. At this point, nothing can be proven and there are many conflicting and divergent stories on record.

Today a place marker stands at the sight of the massacre, located south of US 6 and 50 on an unmarked road. To get there, take a left and head south on the first road (less than a mile) west the intersection of 2500 South and US 6 and 50, just out of Hinckley. The monument is located along the Sevier River at the end of this roughly 2-mile road. Looking at a map will clarify these directions.

Back on the main route (US 6 and 50), resume the westward drive. Check for rations, as Hinckley truly is the last bit of civilization until about 13 miles into Nevada. Not only will there be no gas stations, but there will also be patchier and patchier cell service until there is none at all; if off the main highway, do not expect any type of phone to work, save for a satellite phone.

Leaving the trees and green lawns of town, the landscape quickly becomes more moonlike. The first mountain range passed is the **Cricket Mountains,** on the southern side of the road. Beyond this gateway sits **Sevier Lake,** completely dry—though a mirage often makes it appear to be full of water. Situated between the Cricket Mountains, and the **House Range** (to the west), this massive lake has actually been full recently enough to be witnessed by early settlers—who claimed it supplied water for their farms, as well as for the indigenous Pahvant Indians.

The **House Range,** which runs north-south on both sides of the road, is impressively large—yet bizarre in appearance because of the very sparse vegetation that speckles it. Scattered throughout the hills are old mines, completely invisible from the road. This range has a fair bit of quirky history owing to its mining, hermit, and shepherd activity over the years. It also includes **Notch Peak,** famous for its northwest limestone face, which has one of the tallest, sheer cliffs in the nation at roughly 3,000 feet in height. The peak has an elevation of 9,654 feet, so in total it stands roughly 4,400 feet taller than **Tule Valley,** below.

SPACE!

While driving into the desert, keep scanning to the left and the right for what looks like misplaced double beds in the desert. This "furniture" is actually part of a multimillion dollar international undertaking called **Telescope Array,** a collaborative project among institutions and universities in Japan, Russia, Korea, Taiwan, China, and the United States. The experiment is designed to study cosmic rays, which are high-energy particles entering the earth's atmosphere from space. Cosmic rays originate from events on our own sun, as well as from other, yet unknown happenings on the far reaches of the universe. Until the advent of nuclear testing in the 1950s, cosmic rays had maintained a constant level of Carbon 14 (used in carbon dating) in the Earth's atmosphere. The rays have a number of affects on the earth's atmospheric composition, as well implications in their extraterrestrial creations. Because of this, cosmic rays have been the subject of many studies worldwide since their discovery by Victor Hess in 1912 (who later won the Nobel Prize in Physics 1936 for his discovery of cosmic radiation).

The Telescope Array Project aims to measure these high-energy particles, and its observation field is set up in none other than Millard County, Utah. Five-hundred and seventy-six particle counters (the bedlike objects on either side of the road) have been placed in a 1.2-km-square array. Three telescopes look out into space from the grid. Together these are used to gather the most information possible about these particles—electron,

Cosmic Ray Center in Delta

positron, gamma, muon and more—entering the atmosphere. The telescopes can "see" the florescence created by the particles entering the air; the counters observe of the showers' footprints.

A small office in Delta (right on the northern side of US 6) is the local headquarters and research facility for this, where faculty and graduate students from the participating nations process data and maintain the particle counters. Plans to create a more user-friendly visitors center in Delta are in the works—but meanwhile visitors are encouraged by project supervisors to stop in and check out the work behind the scenes.

Approaching the House Range from the East

While a dozen or so technical routes ascend the cliff face of Notch Peak, non-climbers can enjoy the mountain as well. Just past Sevier Lake's westernmost ledge, look for **Notch Peak Loop Road** to the north. This may or may not be signed; anyone planning to visit the backcountry in this area should bring as detailed a map as possible, with a gazetteer being the minimum standard in detail. Indeed a loop, this road heads north, eventually turning west, and then south again to return to US 6 and 50. However, this is where the map comes in handy: this "loop" is hardly a clean, well signed route. Many, many roads head in and out of the desert, and few are signed. First-hand experience assures this.

Surprisingly, the nontechnical hike to **Notch Peak's summit** takes but a few hours, and the views from atop the cliff give flight to everyone's stomach butterflies. Nearly a mile of airspace beneath the feet is quite dizzying. To get there, take this eastern portion of Notch Peak Loop Road, heading west toward **Sawtooth Canyon.** Park where the road becomes impassible, and begin by walking up a dirt road. In about half a mile, follow the fork to the left, and follow gulches and the mountain shoulder until reaching the top of Notch Peak from a southeasterly direction. Once at the top, the brave can peek over the edge for rock climbing anchors at the top of the face. Again, a detailed map will be extremely useful in the entire off-pavement

CROTALUS OREGANUS LUTOSUS

Common name: Great Basin rattlesnake. When hiking, be aware that this sub-species of western rattlesnake makes its home in western Utah. It lives in grasslands, desert, and forest. Though it typically hangs out on the ground, it will occasionally climb a bush or tree. Fortunately, these snakes prefer to avoid humans, eating mainly smaller dishes like birds, lizards, and mice. They hibernate during the cold months in sheltered areas: small holes and dens. In spring, they begin emerging, seeking sunlight to warm their cold-blooded bodies.

Typically they reach a maximum length of 3 or more feet and have, much like the dry desert dirt, light gray or yellowish tan skin with darker oval spots on the back, as well as a viper head (the jaw of which is significantly wider than the body). Though these aid with recognition, the tail's rattle is the number-one clue. When threatened, these snakes coil up and shake their tail quite rapidly, sounding like baby's rattle on overdrive. They prefer not to encounter people, but when forced they will strike, injecting a potent venom especially undesirable to have in the body whilst in the backcountry.

If bitten, the best thing to do is stay as calm as possible. Adrenaline and extra movements only encourage blood flow to spread the venom. Some simple actions help to reduce the venom's ill effects. It is best to keep the bitten limb beneath the heart and remove any clothing, shoes, rings, or bracelets near the bite, as the limb will swell significantly. If not on a joint, tie a band on top of and below the bite, but not too tight. Wash the wound if supplies are present. If a walk-out is required, it is best to wait as much as a half an hour to allow the venom to localize. Meanwhile, try to memorize the look of the snake in detail, as antivenin is snake-specific. In any case, victims should not cut open the wound or try to drain or suck the venom out; this will not work and will only cause more tissue trauma, introduce bacteria, and increase circulation to the area. Also, tourniquets only increase the likelihood of limb loss.

View into the Chasm in Front of Notch Peak's Cliffs

task, as no roads off US 6 and 50 are paved, the there are no official trails.

Instead of taking the entire Notch Peak Loop, those on the driving detour should turn west (south of the loop's apex) and enter **Marjum Canyon.** Heading into this gap is entering into superb solitude. Even more extreme, imagine living here alone for 20 years, as Bob Stinson did from the mid 1920s until 1946. Back in this day, US 6 and 50 did not exist, and Marjum Pass was the only option for east-west travel through the area. After his service in World War I, legend has it that Stinson returned to Delta to find his wife had married another man. Destroyed, he moved to the canyon, found a natural spring, and built a small cliff dwelling—which is still largely intact today. He used the spring water to keep a garden and even make moonshine. Here he worked maintaining the road, clearing rock slides and keeping travelers company. Eventually US 6 and 50 was built and Stinson, now out of work, moved back to Delta. This road's surface is made up almost entirely of coarse, sharp limestone rocks and should be taken slowly.

To the south of Notch Peak, across US 6 and 50, extends **Tule Valley.** To access this region, keep an eye out for a left-hand turn-off just west of **Skull Rock Pass.** As with most other parts of this "anarchy land," Tule Valley can be accessed via numerous dirt roads, but the main turnoff is the first in this case. A few miles of southern travel on this bring explorers to the eastern edge of a dried-up lake bed. Much smaller than Sevier Lake, this hardpan still stretches more than 5 miles north to south, and more than a mile east to west.

Driving west across the hardpan leads to the central camping area for what is known as **Ibex Wells.** Completely rugged camping, this valley has absolutely no amenities, rangers, fees, or seemingly laws. A fantastically beautiful lunar landscape, this also has nationally famous bouldering and some multipitch sport and traditional rock climbing routes. Quartzite, the rock type here, is incredibly compact and strangely aesthetic—globular and splotchy. Of course quartz, found in quartzite, is a valuable mineral with numerous uses. Farther south is **Topus Mountain,** where a massive quartz mining operation is in full effect. As recently as 10 years ago, this mountain was fully intact; today this obviously no longer the case. Do NOT drive across this or any hardpan if it is raining, has rained recently, or will rain. This surface may look dry, but can hide vehicle-eating mud underneath it. If there is any significant moisture, any car will become stuck and require an expensive extraction by military vehicle.

The last major mountains standing before Nevada belong to the **Confusion Range,** running parallel to, and immediately west of, the House Range. As in the House Mountains, a large network of wild roads trace the areas of the range topographically possible to drive. Numerous small sets of hills, valleys, and deserts lie farther west toward Nevada. **Snake Valley** is the largest of these, straddling the Nevada border.

Cross over this line, into the other state, drive another 3 miles, and head south toward Baker on NV 487, about 5 miles farther along. About a mile before town, look for signs indicating the **Baker Fremont Indian Archaeological Site.** Excavated in the early 1990s, this **Fremont Indian village** is roughly 900 years old. A collection of pit houses and adobe buildings—15 in total—has stored clues for researchers. Here the story told is that of a people that gathered and stored local vegetables, traded minerals and stones with other groups, and possibly even irrigated for agriculture.

The village has an organized, well planned layout uncharacteristic of most Fremont settlements. Unfortunately for viewers today, the sites have

DANITES

Formed in **Far West, Missouri,** in June 1838—about 10 years before the Mormons would even arrive in Utah—the **Danites** were a vigilante fraternal group of Latter-day Saints. At this time, Far West (in present-day Caldwell County) was the headquarters of the Mormon Church. Joseph Smith preached that nearby Jackson County was actually the Garden of Eden, and when Adam and Eve were banished, they came to Caldwell County.

The peace wouldn't last long here for Smith and his group. Just like at every stop along the way, the Mormons found conflict with their neighbors. In the early summer of 1838, several escalating skirmishes took place between Missourians and Mormons. That June, the Danites formed as a secret society and were the key force in the Mormon-instigated attacks. In reaction, the Missouri governor commissioned a 2,500-man army to end the Mormon rebellion. In October the Mormons surrendered and were forced to leave the state.

However, the end of the **1938 Mormon War** did not end militia in the Latter-day Saints. Other body guards of church leaders and protective armies were given different names, such as "Avenging Angels." But many believe that this secret brotherhood in fact moved westward with the church, carrying out cruel attacks on groups the Mormons disliked.

been retro-filled with the same dirt that was excavated in order to preserve the structures' remains for posterity and future studies. Markers have been placed on top of the covered buildings to suggest their subterranean layout. Recovered artifacts are now held at the Brigham Young University Office of Public Archeology Museum in Provo.

From here, drive just more than a mile into **Baker.** This town sits in White Pine County and has fewer than 400 residents. This settlement grew up out of ranching, beginning in the 1870s. Today this town serves primarily as the gate to Great Basin National Park, the businesses are small, often kitschy, and have seasonal hours.

Great Basin National Park definitely stands apart from the other ranges on this drive—looking almost like a misplaced Glacier National Park. Huge mountains rise dominantly from the valley floor, out of the dry desert, into evergreen forest, and finally out of vegetation altogether, with exposed, rocky peaks. A vertical transition of several thousand feet accomplishes more than a thousand miles' worth of travel, from Nevada desert, to Idaho-looking forests, to Canadian Rockies peaks. Massive cirques with sheer, towering cliffs, steep talus fields (atop which often lay snowfields), forests, and lakes. Marked by immense girth and spectacular vertical relief, these mountains rise from the valleys below, more than 1.5 vertical miles, to a maximum height of 13,063 feet at **Wheeler Peak.**

Though the rock shares similar origins as that in the House and Confusion Ranges, it has here been sculpted into massive turrets and horns— a striking view above the green forests. In addition to spectacular mountain scenery, this park is the location of caves and natural arches. Here also grow bristlecone pines, some still living at more than 5,000 years of age.

To get inside the park, take NV 486 west from Baker. This main road leads directly to the **Lehman Caves** less than a mile beyond the park border. These caves may only be entered on a guided tour. Visitors may choose between two different tours (the **Lodge Room** or **Grand Palace** options); each occurs multiple times a day, with 11 total daily tours in summer and four total daily tours in winter. Cave temperature remains constant at 50 degrees F, so dress accordingly. Tickets are priced per person per tour (with varying rates) and can be purchased in person or in advance by calling the park (Monday–Friday only).

After the cave, retrace the drive just a few hundred yards and head northwest on the 12-mile **Wheeler Peak Scenic Drive.** Topping out at more than 10,000 feet above sea level, this road tours up and out of the dry, inhospitable desert into an island of abundant plant and animal life. Rising from about 7,000 feet above sea level at the park boundary, it passes through sagebrush, piñon pine, curleaf mountain mahogany, and into evergreens like ponderosa pine and Douglas fir—not unlike those in the forests of Montana—and finally up into aspen groves at an elevation of 10,000 feet.

Such a dramatic elevation rise causes rapidly cooling temperatures, with plant populations changing completely in as little as 500 vertical feet ascension. Climes found higher select different species into existence, and allow moisture to remain in the soil longer. Additionally, any range so prominent squeezes precipitation from the sky, gathering extra moisture for its inhabitants.

Great Basin National Park, being such a vast wilderness expanse, offers seasons' worth of outdoor activities. Biking is allowed only on pavement within the park; that said, the Wheeler Peak scenic drive should provide a relatively safe, very challenging road bike ride. Abiding by Nevada's conservation laws, visitors can fish any number of creeks in the park. Rock climbing is permitted, too, but climbers must register with the park.

Great Basin National Park marks the end of this route and leaves visitors pretty far out there. To return to the Salt Lake City area, consider driving a bit farther west to Ely, Nevada, and then North to West Wendover. Along the way are two sizeable sections of the **Humboldt-Toiyabe Mountains** and **National Forest,** a beautiful, craggy range with high rocky peaks and glacial lakes, as well as primitive and developed campgrounds. Though most people overlook Nevada, any who have explored it would agree that its mountain ranges rising high above the desert provide amazing backcountry playgrounds similar to those in Montana, Idaho, or Wyoming. Plus, returning to Salt Lake City this way links into the **West Desert: Bonneville Salt Flats Route** (as described in chapter 15 of this book), from West Wendover to Salt Lake City.

IN THE AREA

Attractions and Recreation

Baker Fremont Indian Archaeological Site, Great Basin National Heritage Area, Baker, NV. Located 1 mile north of Baker on NV 487. Call 775-234-7171. Web site: www.greatbasinheritage.org.

Great Basin Historical Society Museum, 328 West 100 North, Delta. Open Monday–Saturday; closed Sunday. Free entry. Call 435-864-5013. Web site: www.millardcounty.com/placestosee/greatbasinmuseum.htm.

Great Basin National Park, 100 Great Basin National Park, Baker, NV. No entrance fee; fees charged to tour Lehman Caves and camp. Call 775-234-7331.Web site: www.nps.gov/grba.

Dining

Rico Taco Shop, 396 West Main Street, Delta. Open daily, breakfast–dinner. Call 435-864-3141.

Other Contacts

Great Basin Business & Tourism Council. Online tourist information. Web site: www.greatbasinpark.com.

Humboldt-Toiyabe National Forest Office, 1200 Franklin Way, Sparks, NV. Call for maps and information; consider ordering maps from home before departure. Call 775-331-6444. Web site: www.fs.fed.us/r4 /htnf.

Telescope Array Project, University of Utah Millard County Cosmic Ray Center, 648 West Main Street, Delta. Visitors encouraged to visit Delta station; call ahead to be sure they're open. Official visitor center in the works. Call University of Utah Cosmic Ray Group, 801-581-6628 or Delta Office, 435-864-1800.

I-80 Bonneville Rest Stop with Views over Speedway

CHAPTER

15

West Desert Salt Flats and the Bonneville Speedway

Estimated length: 120 miles from West Wendover, Nevada, to Salt Lake City, Utah; 175 miles with the Silver Island Mountain Byway detour

Estimated time: Half a day with stops

Getting There: Located on the Utah-Nevada border, right on I-80, no city could be simpler to reach than West Wendover, Nevada.

Highlights: This is a tour of the **Great Basin Desert**—a unique part of the Western United States with no outlet to the ocean. This wicked dry place is saturated with history, ranging from the Donner-Reed Party's crossing, to WWII bomber testing, archaeological sites containing 11,000 years of human history, and the setting of world land-speed records. During various ice ages, the **Salt Lake Valley** and beyond was filled with an enormous body of water, **Lake Bonneville.** Today, the **Great Salt Lake** is just a diminutive memory of that—though plenty of evidence remains. On this drive, keep an eye out for the still visible, ancient shoreline on the mountainsides—several hundred feet higher than the current valley floor. Imagine, too, that the **Bonneville Salt Flats** were once also at the bottom of this massive lake. This lonely drive is almost completely devoid of humanity but tours one of the most unique and desolate landscapes of the Western United States.

Bonneville Speedway

Conspicuously located just west from the Utah-Nevada border, **West Wendover,** Nevada, is a small town with Vegas-like lights. Despite its isolation, West Wendover actually has quite a hold on its neighbors in their socially restrictive state. A shuttle bus runs between Salt Lake City and West Wendover, bringing gambling-deprived and liquor-law-repressed Utahans to natural-resource-deprived border town Nevada. As hokey as this town would seem, many amuse themselves quite genuinely in West Wendover, with its ample casinos, hotels, gas stations, and liquor stores. Given its flashiness, it may come as a surprise that this town has fewer than 5,000 residents.

Heading east into Utah and toward Salt Lake City, travelers pass this town's Utah sister, **Wendover.** This community seems to just be the depressing tailings of its Nevada sibling—for what's a lonely desert town without gambling? Don't scoff just yet; in and around Wendover are natural and manmade points of history that can be appreciated by the prepared, including the Silver Island Mountain Byway, Danger Cave archaeological site, Bonneville Speedway, and Historic Wendover Airfield.

The longest detour departing from Wendover is a 54-mile scenic byway loop, the **Silver Island Mountain National Backcountry Byway.** This road tours along the Silver Island Mountains (surprise!), named such as they truly appear to be an island in the expansive, shimmery Great Basin salt flats. Shaped like an 8 this double-loop route provides an excursion into the wilder parts of these flatlands. Though the byway has numerous entrance and exit points, the easiest access is via I-80's exit 4 (for Bonneville Speedway). When the road makes a 90-degree bend east for the speedway, stay straight, heading slightly left (and away from the speedway) to enter this range. Take a look at a map before heading in; though there is no singular attraction for this route, it allows for a deeper, off-interstate exploration of the incredible salt flats of western Utah. Overland hikes into **Silver, Donner, Lost,** and **Cave canyons** are possible; check online for a map of this route (www.byways.org/explore/byways/68994).

At the apex of this 8-shaped-byway is **Donner-Reed Pass,** part of the **California National Historic Trail.** Though not *the* famous Donner Pass of the Sierra Nevada, this was nevertheless was a stretch of that group's westward journey. This ill-fated trip consisted of roughly 500 wagons attempting to travel from Independence, Missouri to California. They followed the well traveled Oregon Trail westward until Fort Bridger, Wyoming. There, they departed to the south and headed across Utah, a difficult route called Hastings Cutoff. Along the way, numerous obstacles resulted in the loss of life, wagons, and livestock, and created divisions within the group. Making it all the way through the Wasatch, across the Bonneville Salt Flats, they made it as far as the Sierra Nevada in California—quite near modern day Truckee and Lake Tahoe—where tragedy struck in the winter of 1846–1847. Eighty-seven people had arrived in the Sierra Nevada in December 1846. But stranded by severe winter storms, only 48 remained by February 1847, when rescue parties could finally reach the group. During this time, many of the survivors had resorted to cannibalism.

Off the same exit (I-80's Exit 4) is **Danger Cave,** a national historic landmark since 1961. This site is actually a multiple-cave archaeological site containing evidence of human presence dating as far back as 9500 B.C.—

View over Desert from Climbing Cave Near Wendover

more than 11,500 years ago. Ancient peoples from the Desert Great Basin cultures lived in the area for an estimated 10,000 years, first visiting these caves at the conclusion of the last ice age. Various stone tools and weapons from hunter-gatherer cultures have been found in the cave, including knives, drills, and arrow points. Nearly 70 plant species, various animal bones, and basket remnants have also been recovered from the site.

When he excavated these caves in the late 1940s and early 1950s, Jesse Jennings set a new standard for meticulous digging and record taking. His research surprised peers with some of the oldest samplings found in all of North America. Evidence uncovered in the caves suggests that these people were concerned only with survival; gathering food and water apparently was all-consuming, and structure-building, rituals, and art weren't in the time budget. Here, too, some of the oldest pinworm larvae have been found in fossilized people poop, or "coprolite." Dating back to 7800 B.C., this has helped to trace this parasite's path around the world and its co-evolution with humans.

Apparently this was once designated as a Utah state park, but not enough money was put aside and today it remains an undeveloped site. To get there, take exit 4 from I-80. Turn left (north) and a quick left again (west) at the first available turn. This paved road turns to dirt and continues roughly another 1.5 miles and ends at the side of Wendover Mountain. Visible from the road, the caves are located on the eastern side of this

"RESTORE OLD LAKE BONNEVILLE"

A cultish bumper sticker, this is rarely to be spotted in the Utah area anymore. What does it mean to "Restore Old Lake Bonneville"? Partly poking fun at other regional bumper stickers like "Restore Lake Tahoe," this is not an environmentalist cause. It is more like a sick joke about covering many of Utah's northern valleys with hundreds of feet of water.

During the last ice ages—a time when the area's climate was much cooler and wetter, huge lakes formed in the Intermountain West. The Great Basin was a perfect tub for these; with no outlet to the ocean, it grew and shrank with changing climates several times until a natural earthen dam burst 12,000 years ago, draining it once and for all. During the most recent of these fillings, Lake Bonneville, spanned a huge portion of the Great Basin, with shorelines as much as 980 feet higher than those of today's **Great Salt Lake.** Its waters covered an area roughly half the size of present-day Utah (nearly 20,000 square miles). Today, its various shorelines are still visible along the foothills and mountains of the area, with distinct parallel bands marking the different surface levels of the lake throughout its history. Keep an eye out for these across the entire northern portion of the state, from Logan to Tooele, Provo to Tremonton, and beyond.

It is believed that at one point the lake grew large enough to accept the waters of the **Bear River.** Its volume grew too quickly, and the earth at **Red Rock Pass** (near the Utah/Idaho border I-15), suddenly gave way. When this natural dam burst about 12,000 years ago, it unleashed a catastrophic flood that swept westward across southern Idaho through the **Snake River Plane**—a 410-foot-tall wall of water still traveling about 70 miles per hour as it crossed Idaho. Incomprehensible, the flow was estimated to be at 15 million cubic feet per second—more than 1,000 times the flow of the Colorado River's pumping Cataract Canyon.

mountain, with Danger Cave being the northernmost and Juke Box being the southernmost. Visible from the road, these do not require too strenuous a hike. For protective purposes, the entrances have been blocked with iron grates.

Also reached by heading north from I-80's exit 4 is the famous **Bonneville Salt Flats.** World-renowned for the speedway on which land-speed records have been set for more than 100 years, this remarkably flat area has come into existence by a combination of unique factors. As recently as 12,000 years ago, this sat at the bottom of Lake Bonneville, half the size of present-day Utah (see sidebar, above). Part of the Great Basin, a landlocked depression with no outlet to the ocean, this area collects water with no outlet but evaporation. As streams run from surrounding environs into the lake, they carry along certain soluble materials, including various salts. Without a way to drain the water, the lake collects these salts that cannot evaporate. Hence the surface of these flats—salty from the trapped solubles left behind by waters of yore, and almost perfectly flat, as they were once at the bottom of a huge lake.

Over time, an area would tend to get windswept into a less flat expanse—but the varnish-like nature of the salt keeps nature from taking away too much soil. Additionally, each winter a film of water collects on the flats, polishing out any surface irregularities. By spring it evaporates, leaving behind a freshly buffed surface. Plants, too, have been unable to disrupt the surface, as the high-concentration magnesium, lithium, sodium chloride, and potassium make the flats inhospitable for flora.

However, these are not impervious to the powers of nature, and slowly the surface is changing. When measured in 1976, they were roughly 26,000 acres—down from an estimated original expanse of 96,000 acres. Industry, such as massive salt-mining operations (nearly 100 years old) was thought to be responsible for decreasing the surface area and the defacing of the remaining flats. A thinning top layer, as well as the appearance of holes and stress lines, suggests an unnatural disturbance. Regardless of actual cause, this environment should be treated with care. Though Bonneville looks as barren as a piece of concrete, it is in fact fragile, and visitors should remain on proscribed driving paths.

More recently, but as far back as 1896, this was recognized as a possible surface for high-speed racing. W. D. Rishel, who was scouting for a race course for cyclists, first saw this potential. Eighteen years later, he returned with daredevil Teddy Tezlaff. In 1914, Tezlaff drove a Blitzen Benz

Salt at Speedway

141.73 miles per hour across Bonneville for an unofficial record. By the 1930s, land speeds were pushing 280 miles per hour, but were mostly being tested on beaches like Daytona. Needless to say, a flatter, firmer surface would be required to push it much further. In 1935, Bonneville hosted a record at just over 300 miles per hour, sustained for 1 mile. Perfect for land-speed records, these salt flats provide ample space to accelerate to the vehicle's full potential without concern for limitation of the course.

By the late 1940s, the area was the standard course for setting land-speed records. Barriers kept being topped by the hundreds of miles per hour, reaching over 600 miles per hour by the 1960s. Presently, the British "Thrust SSC" holds the record at 763 miles per hour—faster than the speed of sound in some temperature and pressure conditions. This is quite a lot faster than the recorded 1898 record of 39.24 miles per hour. Today (and since the 1930s), the speediest vehicles have all been jet-propelled. There is no one standard for measuring speed—but most records today (and the ones listed above) are the speed a car sustains for 1 full mile.

The World of Speed, a four-day festival for racers and onlookers alike, takes place every year during the middle of September; check the **Utah Salt Flats Racing Association's** Web site (www.saltflats.com) for current listings. Many different categories of vehicles are raced—even the bizarre: racing *bar stools,* which go in excess of 70 miles per hour. Spectators should wear plenty of sunscreen, sunglasses, hats, and also bring lots of hydrating liquids. Ear protection enriches the experience, as do binoculars. There are no grandstands, so bring a camp chair and a radio to catch commentary on AM. The time is long between races, and short during the action, so it's a good idea to come with friends or some other kind of entertainment. To get there, take exit 4 and head north 5 miles, staying on the paved road, to the gate.

To the south of the same exit is the **Historic Wendover Airfield** (sometimes also called the Wendover Field Bomber Training Base or Historic Wendover Air Force Base). In 1940, Wendover, Utah was a town of just more than 100 people, surrounded by miles of uninhabited salt flats. Needing space for a bombing and gunnery range, the U.S. Army Air Corps began acquiring land and erecting structures that would become the Wendover Field. The total size of the field was huge: on average, 27 miles wide and 86 miles long—thought to be the largest of its kind in the world.

The Wendover Army Air Base was officially opened on March 28, 1942, as a training center for B-17 and B-24 heavy bombardment. In all, more than 1,000 crews were trained at the base. Here, too, the Manhattan Project was researched, with "Little Boy" and "Fat Man" being tested on the range. After 155 test drops at this site, and one high-altitude simulation drop in New Mexico, Colonel Tibbets flew a B-29, Enola Gay, on August 6, 1945, to Hiroshima, Japan, to deliver Little Boy. On August 9, Fat Man was dropped by Major Charles Sweeney over Nagasaki. A third bomb (another Fat Man) had already departed from Wendover onboard another B-29 when the Japanese surrendered on August 14. With this news, that plane was intercepted, and returned to Utah.

Today it is possible to arrange for extended tours of the airfield (by calling ahead), and also to visit its museum, open seven days a week. On display are models, military memorabilia, and photographs illustrating the history of the airfield and base. Outside stand the original barracks, shack-like dwellings of the enlisted men during this time. A very powerful history for such a quiet, dreary-seeming place.

Finally leaving this exit 4 and its attractions, head east on I-80. Unmistakably visible 25 miles east of Wendover stands **Metaphor: The Tree of Utah** (commonly called the "Tree of Life"), an 87-foot-tall mixed media sculpture that was built by Karl Momen between 1982 and 1986. This Iran-born Swedish American artist funded the entire project himself, hoping to add color and the appearance of life to the otherwise pristinely stark background. As the story goes, Momen was driving across the desert toward California and envisioned a tree growing there; years later, he realized that vision. Used for the sculpture was 225 tons of concrete, 10,000 pounds of welding rod, roughly 2,000 ceramic tiles, and various rocks and minerals coming from Utah. Apparently there is quite a lot of speculation among passers-by as to what exactly this structure should be. Guesses range from UFO "power pole," to cell tower in disguise, and even an expensive memorial to a wealthy person that perished in a car wreck at the location.

The tree used to stand alone, but vandalism has now forced the State of Utah to encircle it with a barbed-wire-topped chain link fence. Now owned by Utah, the tree is believed to have cost Momen more than one million dollars to complete. On the base of the tree is a plaque enscripted, "A hymn to our universe, whose glory and dimension is beyond all myth and imagination."

Exit 41 is for **Knolls,** Utah. A small town until sometime in the 1970s, Knolls no longer exists. The main attraction for the exit is actually the **Knolls Recreation Area,** so named for the rolling hills and sand dunes of the area. This 36,000-acre playground is beloved by dirtbikers and ATVers. With very sparse vegetation and many natural features, the park is perfect for challenging riding, as well as doing jumps and tricks. Riders can often be seen from the interstate as many popular jumps sit almost adjacent to I-80; sometimes the highway itself is even (illegally) used to gain speed for a jump.

Do not exit for Clive, Utah (the next along the way), unless you're interested in visiting a privately owned nuclear waste site. Clive's claim to fame is having one of few sites in the nation that will accept commercially generated byproducts—the others being in Barnwell, South Carolina, Richland, Washington, and (one pending) in Fort Hancock, Texas. The owner of the Clive site signed an agreement with then governor John Huntsman in

2008, agreeing to only accept Class A nuclear waste (the lowest radioactivity classification), and only a limited amount of it.

Move right along and take exit 56 for a quick tour of the ghost town, **Argonite.** Head south from the interstate about 2 miles, and turn left, continuing about 3.5 miles on a high-quality dirt road. At this point appear abandoned buildings (some of which are still in fairly good condition), mine shafts, and one old vehicle. This village operated as mining town just a handful of years in the early 1900s. It fell quickly out of fashion as its quarried stone, aragonite, had little use but decoration in buildings. With little else to support its economy, this remote town collapsed. Mining resumed decades later, but only under the power of commuting workers; the homes and buildings were never used again. Because it was haphazardly abandoned, this site contains uncovered pits and shafts; pay attention and be careful.

Continuing eastward, glance to the right at Exit 70 at **Delle.** Like Aragonite, this is also a ghost town (or very near to it). Instead of growing up on mining, though, this came to be as a railroad village on the Western Pacific Line. Delle's original name, Dalles Spring, was shortened and slightly altered by railroad workers in the name of easier telegraphing. Families of rail employees lived here for about 20 years, approximately from 1920 to 1940, piping in water from a spring more than 10 miles away. But changing rail technology and the inhospitable salt-flat environment led these families to all but desert this town. Originating in the San Jose/Oakland area of California, the Western Pacific Line scooped north and east through northeastern California, northern Nevada, and across Utah toward Salt Lake City. The shortness of this railroad (only about 1,000 miles on its the main line), and the high-maintenance nature of the mountainous route, only allowed it to be in full-swing operation until the deregulation of railroads in 1980. Unable to compete with major lines in the area, Western Pacific was forced to be purchased by Union Pacific at the end of 1982. Today exists still a small Sinclair gas station, as well as a dusty-looking hotel and café. A de-wheeled, de-everythinged red bus sits eternally parked there as well, covered with various murals.

UT 196 shares Delle's same interstate exit ramp (Exit 77). Taking this road about 37 miles to the south leads to **Dugway,** the gateway to the **Dugway Proving Ground.** This small town of roughly 2,000 people serves a bedroom community for the famous and massive military test site, founded in 1941. Here, the U.S. Army aimed to test chemical and biological weapons in an isolated and deserted environment. Looking around, it's hard to argue their logic; however, scandal has struck the base, as many past scientists and personnel there have spoken out about the potential leaks into the environment, as well as harm experienced by them and regional livestock. In any case, it is believed that more than a thousand different chemicals were tested here. Dirty bombs were said to have been dropped, as well as biological weapons and chemicals released into the air, and deployment of weapons of mass destruction. Needless to say, much of the activity here has been secretive.

The **1968 Skull Valley Sheep Kill,** also known as the Dugway Sheep Accident, is an event in which 6,000 sheep belonging to Skull Valley ranchers (just east of the Dugway Proving Ground) suddenly perished. In addition to the sheeps' gruesome suffering and death, the ranchers suffered $300,000 loss in livestock (which, adjusted for inflation, would be just more than $1.9 million).

On March 13 of 1968, three separate testing operations took place, including the burning of a nerve agent, firing of chemical artillery shell, and spraying of nerve agent by jet plane. Most believe the jet plane test is responsible for this "sheep accident." Based on reports, the jet's spray nozzle malfunctioned, causing and the plane therefore to release toxins into the air as it climbed above a safe altitude.

However, other reports raise a bit of suspicion—at least as to exactly what chemical killed the sheep. Though the sudden mass death of so many thousand grazing animals was clearly not natural, some of the sheep observed to be alive after the incident appeared to be suffering from something other than nerve gas exposure—they were bleeding internally and refused to eat, yet their breathing was normal. Typically nerve gas, which stops the breakdown of neurotransmitters that normally allow muscles to relax, causes (among many horrid symptoms) tightness in the chest, a loss of bodily control, and inability to breathe. Additionally, it seems that only sheep were affected, while other animals reportedly survived. The army did not officially admit guilt for this massacre, but they did bury a few thousand of the dead animals and also replaced much of the lost livestock for the ranchers.

Regardless, President Richard Nixon, reacting in the face of international exposure, banned all open-air testing of chemical weapons the following year. And in a 1970 report (finally made public in 1998), the Army officially admitted that Nerve Agent VX was responsible for the killing of the sheep. In this report, it was said that the agent was indeed found present in grass and snow samplings taken from Skull Valley following the sheep's death. Though the Army conceded that the agent was indeed sufficient to kill the sheep, it still did not claim negligence or responsibility.

An article in *Time Magazine,* "Toxicology: Sheep & the Army," published April 5, 1968, details much of the veterinarians' reports and news conferences before time faded them. It is available to read online (www.time.com/time /magazine/article/0,9171,900098,00.html).

Many of the facts of the Dugway Proving Ground will obviously never be set straight. That said, many might feel spooked out and inclined to pass by as quickly as possible. As one *Houston Chronicle* reporter, Tony Freemantle, said,

> Dugway Proving Ground is a massive firing range that for fifty years was the U.S. Army testing ground for some of the most lethal chemical, biological, and nuclear weapons man ever made. A slope of the mountains to the east is pockmarked with hundreds of fortified bunkers storing enough toxin to eradicate mankind. This was where the Cold War was waged, not on battlefields in foreign lands but in factories and laboratories and testing ranges.

The Registry of Atmospheric Testing Survivors (or RATS) keep a Web site compiling first-hand testimonies, accounts, and relevant news. In addition to details on the Dugway Proving Ground, this group discusses the entire topic of military chemical and biological agent testing, on land and at sea. In a story with such disparate accounts and oft secretive chapters, it's hard to know whom to believe, but this tells an interesting counter story.

Continue eastward on I-80 into **Grantsville** for a visit to the **Donner-Reed Museum.** Unlike many of the mining and railroad towns in the West Desert area, Grantsville was founded as a satellite Mormon community in 1850, under the direction of Brigham Young—shortly after the LDS arrival in Salt Lake City in the summer of 1848. The history of this town was therefore always interlaced with Utah's unique Mormon story.

The actual museum is a small affair, dedicated to telling the story of the famous party that became stranded in the Sierra Nevada in late 1846, some of whose survivors resorted to cannibalizing the dead in order to survive the bitter winter. Prior to their arrival and snowy entrapment in California, they encountered great difficulties on the Hastings Cutoff, a southern spur route that departed from the standard Oregon Trail at Fort Bridger, Wyoming, and cut across Utah. This "shortcut" was much more challenging and vastly less traveled than the standard Oregon Trail, forging through the high Wasatch Range east of Salt Lake City and across the Great Basin salt flats. Hastings Cutoff was roughly parallel to (but much more "squiggly" than) modern I-80. Lansford Hastings, the namesake pioneer guide, was supposed to guide them, but he apparently gave misleading information about the ease of the route and then left them stranded.

At the worst of times, the party had to cut a road in the Wasatch Mountains for miles through willows, only to find they must turn around and return the way they had come. The Great Salt Lake finally became visible to them in late August, and travel across the Great Basin Desert would prove much quicker. Still oxen would die of exhaustion and dehydration crossing these blazing hot salt flats and these dead animals, as well as "extra" wagons, would be left behind. The extreme difficulties of crossing Utah in 1846 slowed the party greatly, causing them to arrive finally in the Sierra Nevada in December of that year—much too late for passage through such a massive range. Trapped by blizzards, the in-trouble party sent members away to request rescue, but they remained unaided until February of 1847.

Today the museum holds the remains of five abandoned wagons, horse shoes, yokes, tools, jewelry, and other miscellaneous items, found in the Salt Lake Desert in the 1920s—preserved remarkably well here, despite utter exposure to the elements, by the arid climate. Though it is possible some of the items were left behind by miners taking the Hastings Cutoff in 1849, these are most likely those left behind by the Donner-Reed Party three years prior.

Tooele is the next major stop along the way. This requires yet another detour from the main drive, but for those interested in speed, the **Miller Motorsports Park** could be a logical destination after visiting the Bonneville Speedway. This race complex has something for nearly every genre of vehicular racing and competition. On the premises is a modern-day, high-speed, paved track used by race cars and knee-dragging motorcyclists. Additionally, short-course, off-road racing events take place off the track once in a while. And there's even an "extreme rock crawling" course in which some vehicles invariably roll in every competition. Contestants span from amateur to professional, but all races are typically quite high-speed and lively—even the amateur events. Plus, race days typically feature a full spectrum of levels—so with one entrance fee, you'll be able to see amateur

Tooele after a Rain Storm

to full-blown pro. To get there, take UT 36 south about 3 miles, turning right on UT 138. Follow this west for almost 5 miles and, guided by signs, bear south on Sheep Lane for another 2 miles.

After Tooele, it's pretty much a straight shot to Salt Lake City without a heck of a lot between. Notice the **Great Salt Lake** to the north; the low-tide-like smell sometimes wafting south should serve as a reminder. On its south shore about 20 miles west of central Salt Lake City, is the **Great Salt Air**, a cool, wharf-like structure that has been built, destroyed by fire, and rebuilt three times in its history since first opening in 1893. Not just a victim of fire, it was also flooded when the Great Salt Lake reached epic heights in 1983. Called "Coney Island of the West," earlier generations of the structure indeed resembled an ocean wharf—absolutely massive and almost spooky in historic photos. In its earliest days, it was a full-blown resort, catering specifically to Mormons who wanted to vacation at a place with a good reputation, high moral standards, and clean activities. Located directly on a train line, it attracted people from around the region. Today this much simplified building functions as a concert venue and is in much less grand a scale than its original format and made of nonflammable concrete.

Arriving in Salt Lake City completes the day's journey. Luckily, Salt Lake City is the state's major hub, with a major airport and two interstates running directly out of it. From Salt Lake City, the nearest drives are through the **Little Cottonwood Canyon, Big Cottonwood Canyon,** and **The Alpine Loop** (detailed in chapters 1, 2, and 3 of this book).

IN THE AREA

Attractions and Recreation

Bonneville Salt Flats, Bureau of Land Management Salt Lake District Office, 2370 South 2300 West, Salt Lake City. Call 801-977-4300. Web site: www.blm.gov.

Bonneville Racing, Bonneville Salt Flats. Check online for racing schedule. Web site: www.bonnevilleracing.com.

Danger Cave Archaeological Site, Utah Statewide Archaeology Society, Salt Lake/Davis County Chapter. Contact Jennifer Elskin at jelsken@swca .com. Web site: www.utaharchaeology.org/sldavis.

Donner-Reed Museum, 90 North Cooley, Grantsville. Open by appointment only; call ahead to schedule a meeting. Visit the Web site for history of the party, particularly the Utah (or "Hastings Cutoff") portion. Call 435-884-0824 or 435-884-3411. Web site: www.donner-reed-museum.org.

Historic Wendover Airfield, 345 South Airport Apron, Wendover. Museum open daily; arrive at least one hour before closing. Call 801-571-2907. Web site: www.wendoverairbase.com.

Great Saltair, Exit 104 on I-80, Magna. Call 801-448-7206. Web site: www.thesaltair.com.

Miller Motorsports Park, 2901 Sheep Lane, Tooele. Call 435-277-7223. Web site: www.millermotorsportspark.com.

Silver Island Mountains Scenic Byway, exit 4 on I-80, UT. Web site: www.byways.org.

Other Contacts

Registry of Atmospheric Testing Survivors. Web site: www.project -112shad-fdn.com.

Utah Salt Flats Racing Association, P.O. Box 27365, Salt Lake City 84127. Organizers for World of Speed; admission charged daily or for entire event. Call 801-485-2662 or 801-583-3765. Web site: www.salt flats.com.